SCIENTIFIC
SOCIALISM

ISBN: 978-1-963068-79-5 (sc)
ISBN: 978-1-963068-80-1 (hc)
ISBN: 978-1-963068-81-8 (e)

Library of Congress Control Number: 2024922784

2024.10.31

CHAPTER 1

CREATE AN INDEPENDENT CONGRESSIONAL ASSOCIATION

The House of Representatives recently managed to avert a federal government shutdown, with a mere few hours to spare. As they could not agree on an annual budget, they compromised, or "kicked the can down the road", with a forty-five day "Continuing Resolution", or "CR". This compromise satisfied absolutely no one!

In fact, "The House", as the press is currently referring to it, which is also known as the Congress, is controlled by the Republican Party, by the slimmest of margins. That Party is referred to as the "GOP", or Grand Old Party, although most Republican Party members no longer think the Party is quite so "Grand"! The Members of that Party are at each others throats!

The "rank and file" Members of the GOP are not at all satisfied with the leaders of the Party. Several of them decided to take action. In public, the leaders of the Republican Party are referring to them as the "Eight Rebels", as "Legislative Terrorists", as "Saboteurs" and "Pyromaniacs".

No doubt, he is well aware that his constituency, the voters within his Congressional District, expect him to represent their best interests. As a person of principle, he went to Washington, as a Member of Congress, determined to do just that. Yet he found that he cannot! Due to forces that are clearly beyond his control!

In his frustration, he refers to certain "foundational problems" within the "conference". It is very likely that the "conference", to which he is referring, is nothing other than the Republican Party, the GOP. It is also very likely that the "foundational problems", is a reference to the policy of that same Republican Party. He thinks the GOP should change its policy! Without a change in policy, he expects another "blow up"!

This Member of Congress even went so far as to compare the Republican Party, of which he is a Member, to a "clown car"! Without a "fundamental change in policy", he is convinced that replacing Kevin McCarthy, as Speaker of the House, with a different Republican, would merely result in a "different driver"! Nothing would change! He is so right!

He also referred to the "pyromaniacs and saboteurs", the "chaos artists", those who crave the "attention and chaos". It is not clear if he is referring to the leaders of the Republican Party, or possibly the Eight Rebels, or perhaps both. As he was speaking from a deep sense of anger and frustration, of bitterness, it is entirely possible that even he is not sure!

That in no way changes the fact that the leaders of the Republican Party, are not "pyromaniacs and saboteurs". Nor are they "chaos artists". On the contrary, they are entirely focused! They are the loyal and devoted servants of the billionaires! The success - or failure! - of the American government, is of no concern to them! The same is true of the Democratic Party!

The fact that a rank and file Member of the GOP has dared to speak out, in such a manner, is an indication of the strength of the revolutionary movement. Within the House of Representatives, at least within the Republican Party, discipline has completely broken down. It is clear that the revolution has reached the politicians in Washington!

CHAPTER 1

CREATE AN INDEPENDENT CONGRESSIONAL ASSOCIATION

The House of Representatives recently managed to avert a federal government shutdown, with a mere few hours to spare. As they could not agree on an annual budget, they compromised, or "kicked the can down the road", with a forty-five day "Continuing Resolution", or "CR". This compromise satisfied absolutely no one!

In fact, "The House", as the press is currently referring to it, which is also known as the Congress, is controlled by the Republican Party, by the slimmest of margins. That Party is referred to as the "GOP", or Grand Old Party, although most Republican Party members no longer think the Party is quite so "Grand"! The Members of that Party are at each others throats!

The "rank and file" Members of the GOP are not at all satisfied with the leaders of the Party. Several of them decided to take action. In public, the leaders of the Republican Party are referring to them as the "Eight Rebels", as "Legislative Terrorists", as "Saboteurs" and "Pyromaniacs".

In private, the Party leaders are referring to them in somewhat stronger terms!

These "Eight Rebels" are of the opinion that the GOP made some "unacceptable concessions" to the Democrats, in order to avoid a government shutdown. For that reason, they were able to force a vote, which resulted in the GOP Leader, Kevin McCarthy, losing his position, as Speaker of the House.

The House no longer has a Speaker! The House is deadlocked! Washington is deadlocked! No bills can be passed - or even considered! - until the House elects a Speaker!

The journalists are reporting that they have never before heard such "political rhetoric"! The Members of the GOP have been meeting "behind closed doors", in an attempt to find a replacement for McCarthy. Some Members want the Eight Rebels kicked out of the Party! Personal insults have been exchanged! Amid "flaring tempers", certain Members have come close to "fist fights"!

In an attempt to "cool raging tempers", the leaders of the GOP decided to adjourn for a week!

As the journalists have pointed out, as soon as the House resumes, they will have a mere five weeks to first elect a new Speaker, and then vote to pass a new budget. Without a Speaker, no such vote can take place, and the federal government will shut down, in mid-November.

On the other hand, even with a Speaker, the Eight Rebels are determined to "extract impossibly large concessions from the White House and the Senate", according to the journalists. Even if the House passes such a budget, it is very likely that the Senate and White House will not agree. For that reason, the journalists are convinced that the government is certain to shut down, in the middle of November.

This "governmental crisis" is being taken quite seriously, at least by certain scholars. In the simplicity of their souls, they consider this to be a "warning sign for democracy".

This is true, as far as it goes! These scholars -all of whom are bourgeois! - are careful to avoid class terms! They are careful to avoid saying that this crisis is a threat to *bourgeois* democracy! The democracy of the capitalists! The democracy of the billionaires!

Democracy is nothing other than a method of *class rule!* It is a method by which one class crushes another class! As we live under capitalism, it is the method by which the monopoly capitalists, the billionaires, crush the "lower classes", mainly the working class, but also the family farmers, and the middle class.

This is the "democracy", the "method of class rule", which is being threatened! This is precisely the little detail which the bourgeois scholars, neglect to mention!

One GOP rank and file Member of the House, stated the problem, as he saw it, in terms which are quite "colourful", as well as entertaining. He is not one of the Eight Rebels who voted to oust McCarthy, as Speaker of the House. Yet he expressed sentiments which closely match that of the "Rebels". For that reason, it is reasonable to assume that his opinion is that of a great many rank and file Republican Members of the House of Representatives:

"If we do not change the foundational problems within our conference, it is just going to be the same stupid clown car with a different driver... the reality is that we are just going to have another blow up.... The pyromaniacs and saboteurs are not going to be satisfied after they burned down one house. I think they are going to have an itching to go burn down a couple more... It is not time for polite talk... We are 33 Trillion dollars in debt, we have a southern border in crisis, and these guys, chaos artists, decide that we need to take a couple of weeks off from governing the country... They crave the attention and chaos..."

This statement is most significant, and not merely because it is entertaining. It clearly expresses the point of view of a number of young politicians, those who were recently elected. This rank and file Republican Member of Congress, is clearly consumed with bitterness and frustration.

No doubt, he is well aware that his constituency, the voters within his Congressional District, expect him to represent their best interests. As a person of principle, he went to Washington, as a Member of Congress, determined to do just that. Yet he found that he cannot! Due to forces that are clearly beyond his control!

In his frustration, he refers to certain "foundational problems" within the "conference". It is very likely that the "conference", to which he is referring, is nothing other than the Republican Party, the GOP. It is also very likely that the "foundational problems", is a reference to the policy of that same Republican Party. He thinks the GOP should change its policy! Without a change in policy, he expects another "blow up"!

This Member of Congress even went so far as to compare the Republican Party, of which he is a Member, to a "clown car"! Without a "fundamental change in policy", he is convinced that replacing Kevin McCarthy, as Speaker of the House, with a different Republican, would merely result in a "different driver"! Nothing would change! He is so right!

He also referred to the "pyromaniacs and saboteurs", the "chaos artists", those who crave the "attention and chaos". It is not clear if he is referring to the leaders of the Republican Party, or possibly the Eight Rebels, or perhaps both. As he was speaking from a deep sense of anger and frustration, of bitterness, it is entirely possible that even he is not sure!

That in no way changes the fact that the leaders of the Republican Party, are not "pyromaniacs and saboteurs". Nor are they "chaos artists". On the contrary, they are entirely focused! They are the loyal and devoted servants of the billionaires! The success - or failure! - of the American government, is of no concern to them! The same is true of the Democratic Party!

The fact that a rank and file Member of the GOP has dared to speak out, in such a manner, is an indication of the strength of the revolutionary movement. Within the House of Representatives, at least within the Republican Party, discipline has completely broken down. It is clear that the revolution has reached the politicians in Washington!

Another individual, who is considered to be the leader of the Eight Rebels, and is thought to be responsible for the removal of Kevin McCarthy, as Speaker of the House, also had a few words to say:

"We are 33 Trillion dollars in debt. We are facing 2.2 Trillion dollar annual deficits. We are facing de dollarization globally. That will crush working class Americans. Kevin McCarthy is a creature of the swamp. He has risen to power by collecting special interest money and re-distributing that money in exchange for favours. We are breaking the fever now. We should elect a Speaker who is better. What is paralyzing the House of Representatives was not taking up appropriations Bills. We lost six weeks vacation while the appropriations process hung in the balance... The House of Representatives has been paralyzed for the last several decades as we refused to pass a budget and was governed by Continuing Resolutions and Omnibus Bills. I think this represents the ripping off of the band aid and that is what we need to do to get back on track..."

This statement, by another rank and file member of Congress, is also quite remarkable. It is clear that he too, speaks from a deep sense of anger and frustration. His accusations, against the former Speaker of the House, are very serious, possibly even slanderous. But as a Member of Congress, he can perhaps get away with this.

Yet he is correct that the House has consistently refused to "pass a budget", instead choosing to "kick the can down the road", with "Continuing Resolutions" and "Omnibus Bills". He is also correct when he stated that the country is "33 Trillion dollars in debt", with "2.2 Trillion dollar annual deficits".

He stated the problem correctly. His solution, is that of voting out Kevin McCarthy as Speaker. Then he proposes to "elect a Speaker who is better". He thinks that is what is needed to "get back on track"!

Such starry eyed optimism is most touching! And completely out of place! He is a Member of the House of Representatives! Yet he sounds like a Boy Scout! Perhaps he thinks his "good deed for the day" was voting to remove Kevin McCarthy as Speaker!

Reality check! There can be no question of getting the House "back on track", because the House is completely "on track"! The leaders of the House are the loyal and devoted servants of the billionaires! The House has *not* been "paralyzed for the last several decades"! On the contrary, the House has been working very hard, in the service of the billionaires! No change of face of the Speaker, is going to change that!

That bitter lesson was learned recently, by those who took part in the Occupy Movement! At the start of the Movement, common people thought that the system was not working properly. So they decided to exercise their democratic right to peaceful protest. They actually expected their democratically elected leaders to listen to them. The response of the government, that of clubs, pepper spray and tearing down of tents, handcuffs and jail, made it quite clear to those protesters, that the system is working precisely the way it is supposed to work! In the service of the billionaires!

That was a different time, and a different revolutionary movement. Our current revolutionary movement is much stronger and deeper! It has even penetrated the "hallowed halls" of the House of Representatives!

This can best be explained by Lenin, in his excellent work, Left Wing Communism, An Infantile Disorder: "Symptomatic of any genuine revolution is a rapid, tenfold and even hundredfold increase in the size of the working and oppressed masses -hitherto apathetic- who are capable of waging the political struggle".

As a result of this "genuine revolution", countless people, those who were "hitherto apathetic", are now "waging the political struggle". They have pressured the Party bosses, of both Democrats and Republicans, into nominating people of principle, for political office. A number of these people of principle were elected and sent to Washington, especially to the House of Representatives, the Congress.

As these newly elected Members of Congress are people of principle, they set out to do that which they were elected to do. To govern! To represent their constituency! To do that which is in the best interests of the people who voted them into office! To perform their sworn duty! To "preserve, protect and defend the Constitution"! This duty involves defying the bosses, of both Parties!

These principled Members of Congress deserve our whole hearted admiration and support! They are very likely rather young, and expected something entirely different! They believed the lies they had been told all their lives! They actually thought that now that they were elected to Congress, they could go to Washington and make a difference! Fat chance!

This explains their frustration and anger, their bitterness! Although this is not pleasant, there is no harm in this. It is all part of the process of becoming tempered. It cannot be avoided. They will emerge from this class struggle far stronger! Veterans of the class war!

With that in mind, may I suggest that seasoned veterans, perhaps of the Occupy Movement, get in touch with these people. Share with them your experience. Just state the facts.

The fact is that at the beginning of the Occupy Movement, countless people were concerned that the government was not properly representing the common people. It seemed to be favouring the billionaires, those who were referred to as the "1%". Everyone thought that the democratically elected leaders, the politicians, were not aware of this. So people decided to protest, as a means of advising those leaders, those "servants of the people".

The protesters actually thought that, once those leaders were aware of this, that the government officials would change their policy! The protesters expected the politicians to quit favouring the billionaires, and start treating all citizens equally! Ha!

In all fairness to the politicians, they did respond to those legitimate grievances! The police were turned loose! Like mad dogs, they attacked American citizens, those who were exercising their Constitutional right to peaceful protest!

Those protesters were at first devastated! Consumed with anger and frustration! Supremely bitter! To think that they have been lied to all of their lives! To find out that the people whom they - formerly! - admired and respected, are a pack of liars and hypocrites!

Now that the revolutionary motion is raging, once again, a whole new generation of common people have to go through the process of becoming tempered. It is not pleasant! Yet it cannot be helped! It is absolutely necessary! There is no harm in this. In no other way can people become veterans!

Soon, this new generation will return to battle, stronger than ever! They will know what to expect! It is not a school yard squabble! It is a battle of survival! The government is in the service of the billionaires! The billionaires are determined to become trillionaires! At our expense! Class warfare!

The billionaires are in charge! The government is set up to protect the billionaires. Nothing else! The suffering of the common people is of no concern to them! Sentiment has no monetary value! Whether the government shuts down, or even goes broke, is a matter of complete indifference, to the billionaires!

It is the system that has to be destroyed, as it cannot be "patched up"! That calls for revolution. Those who are placed in positions of authority, in Washington, such as in the House, can play a key role in that revolution!

For such people, at the very least, a careful reading of State and Revolution, by Lenin, is necessary.

May I suggest that our veterans, of the Occupy Movement, provide the newly elected, principled Members of Congress, with a word of advice. Feel free to face the fact that, at the time of the revolution, the existing state apparatus is certain to be destroyed. That is a fundamental tenet of Marxism. It is then to be replaced with a new state apparatus, in the form of the Dictatorship of the Proletariat. This is referred to as Scientific Socialism.

As that is the case, it is imperative that all such Members make quite clear, their support for the revolution. Otherwise, they could become a target of that revolution!

The experience of previous revolutions has proven, that it is to the advantage of the working class, to have politicians who support the revolution, within the Capital. Those politicians do not have to be Communists, or even self-proclaimed socialists. But it is important that they support the revolution.

There are a couple facts which should be stressed. One of those facts is that the only alternative to capitalism is Scientific Socialism. The second fact is that the billionaires have just made it clear that all banks but eight, and all businesses but five, are "Too Small to Succeed".

As all principled Members of Congress are almost certainly middle class, this just means that they are facing financial ruin. It is in their best interests to support the revolution. After the revolution, under the Dictatorship of the Proletariat, their training and skills will prove to be most valuable. They can expect to be rewarded.

That in no way changes the fact that all such Members are under extreme pressure, from their respective Party Leaders. Just as the GOP Members are under pressure from their Party bosses, so too the Democratic Members of Congress, are also under such pressure from their Party bosses.

Yet there is strength in numbers! All union workers can testify to that! Not that it is reasonable to expect middle class Members of Congress to form a union! But perhaps an Independent Association of Members of Congress! The rank and file GOP Members can "reach across the aisle", to their Democratic colleagues! All have taken the same oath, to "preserve, protect and defend the Constitution". That oath can form the foundation of their Association!

For that matter, feel free to reach "across the chamber", to the Senate! It is very likely that many rank and file Senators may be anxious to join the Independent Association of Members of Congress!

Bear in mind that, even after the Great October Socialist Russian Revolution of 1917, the Russian Duma continued to function. It is entirely possible that after the American Revolution, the Congress will also continue to function. Not that it is likely that both the House and the Senate will continue in operation. After all, a country needs only one legislative body. That remains to be seen.

No doubt, voices will be raised, in objection to Members of Congress, from both Parties, coming together to form an Independent Association. True, each Member ran for office as a Member of one or the other mainstream political party, either Republican or Democratic. For that reason, there are those who maintain that they have a moral obligation to be loyal to that particular party.

In response to this legitimate concern, our principled Members of Congress can respond that their main responsibility is towards the Constitution! They took an oath to "preserve, protect and defend the Constitution!" The *Constitution, not the Party!*

As the leaders, of both Parties, are working in the service of the billionaires, it is simply not possible for principled Members of Congress, to remain loyal to their Party! Conscience dictates that they form an Independent Association of Members of Congress. That is the only way in which they can properly serve their constituency!

The more rank and file Members of Congress who join the Association, the better! If they form the majority, they can then elect a new Speaker of the House! The newly united House can then pass a balanced budget, one that requires the billionaires to pay their fair share of taxes.

Of course, such a budget would also have to pass the Senate, before going to the White House, for the president to sign into law. Yet if enough Senators join the Independent Association of Members of Congress, that is not a problem! They could even have enough votes to overturn a presidential veto!

As well, the government will continue to function, avoiding a shutdown. Problem solved!

In conclusion, we can say that the revolutionary movement is now so strong, it has reached into the "hallowed halls" of the House of Representatives, possibly also the Senate.

It is entirely possible, that the only way in which a new Speaker of the House can be elected, is through the creation of an Independent Association of Members of Congress.

The one week "cooling off period" could well backfire! The ranks of the "Rebels" could swell considerably! The efforts of the Party bosses, to restore "law and order" within the ranks, could well force the Members to band together in protest! To form an Association!

Without doubt, the approaching few weeks are bound to be most interesting. As the common people are watching events closely, it is bound to raise their level of awareness.

It is the duty of Communists, to get in touch with as many progressive rank and file Members of Congress as possible, and offer them our full support! As well, get in touch with as many common people as possible, if only through the internet. Explain to them that which is happening. Encourage people to read State and Revolution, at the very minimum.

We have no way of knowing how this is going to play out. We do know that the revolutionary motion has touched down in Washington! This is most encouraging.

At the time Congress resumes, it is doubtful that the Party bosses will be able to elect a new Speaker. It is entirely possible that the rank and file Members, will relieve them of that burden! They may elect their own Speaker! And then pass an annual budget, one which taxes the billionaires! As well as avoid a government shut down!

Or perhaps the Members of the House will continue to fight among themselves, unable to elect a Speaker, unable to avoid a government shutdown.

The important thing is that the revolution has reached Washington, and the working people are paying close attention.

CHAPTER 2

CONCERNING THE ISRAEL - HAMAS WAR

On October 7, 2023, Hamas went to war with Israel. In fact, they mounted a major offensive, attacking by land, sea and air, from the Gaza Strip. Israelis are referring to this as their "Pearl Harbour", or their "Nine Eleven", as the surprise was so complete, and the attack was so effective. They were snapped out of their complacency! There were a great many Israeli casualties, civilian as well as military.

The offensive was supremely well planned, in complete secrecy, over a period of perhaps several years. The Israelis thought that the Gaza Strip was safely contained, as it is only forty-one kilometers long, and twelve kilometers wide, at best. It has a total area of three hundred fifty square kilometers. With a population of over two million people, it is one of the most densely populated areas in the world.

The Israelis built a most impressive wall, around this small strip of land, one which opens up onto the Mediterranean Sea. This Wall is six meters tall, or about twenty feet. There are also underground concrete barriers, to prevent tunneling. It is fitted with hundreds of cameras, towers, radar, sensors, and guards stationed every one hundred fifty meters, or five hundred feet. It was thought to be a model of high tech, and impenetrable! They could not possibly have been more mistaken!

This "high tech" defense proved to be about as effective, as the French Maginot Line! That Line was meant to stop the German offensive in World War Two! The Germans breached that Line with ease! Hamas just as easily, breached the Israeli Wall!

On that fateful day, using commercially available drones, Hamas bombed the towers and communication infrastructure, so that the Israeli Defense Force, the IDF, was effectively blinded. Then a wave of paragliders flew over the Wall. Bulldozers opened the Wall in numerous places, allowing bigger vehicles to drive through. The supposedly impenetrable Wall was pierced in no less that twenty-nine positions, at the same time! Hundreds of Hamas fighters poured through, attacking Israel!

A great many IDF soldiers and civilians were killed. As well, numerous "hostages" were captured, and taken into Gaza. As yet, no one knows the precise number.

At the same time, from within Gaza, a great many rockets were fired at Israel. Over a period of several days, perhaps five thousand rockets! Some of these rockets had a range of seventy kilometers or forty miles! Israel reports that most of them were shot down by their missile defense shield, referred to as "Iron Dome". On the other hand, impartial observers report that most of the missiles hit their mark!

Israel responded with "air strikes" against Hamas strong points, within Gaza. As it is so densely populated, this caused countless civilian casualties. They have also placed Gaza under a "state of siege". Nothing and no one is allowed in or out. The electric power has been cut off. No food, water or medical supplies are allowed into Gaza. The fishers are not allowed to go fishing.

In fact, Doctors Without Borders are describing conditions within Gaza as "catastrophic". The hospitals are currently running on generators, and will soon run out of fuel. Those same hospitals are already low on supplies, and without electricity, they "risk turning into morgues"!

The International Committee of the Red Cross said on October 12, that it "stood ready to help". As they stated, "We have people on the ground, and we are ready to play our role of neutral intermediator and contribute to release of these people", by whom they meant the hostage.

As well, UNICEF is preparing to send humanitarian assistance to Gaza, through the Rafah border crossing, between Egypt and Gaza. Aid agencies in Egypt are preparing to send urgently needed humanitarian supplies into Gaza, and receive wounded Palestinians.

The Israeli energy minister made it quite clear that he is not interested. He is determined that there will be "no pause in the siege" until Hamas releases the hostages in Gaza. As he stated, "Humanitarian aid to Gaza? No electrical switch will be lifted, no water hydrant will be opened, and no fuel truck will enter until the Israeli hostages are returned home. Humanitarian for humanitarian. And nobody should preach us morals".

This tactic of "total siege" is out lawed under international law! It amounts to declaring war on civilians! A crime against humanity! It necessarily leads to a humanitarian disaster! Yet the western powers are still supporting the Israelis!

The American Secretary of State made that most clear: "The message that I bring to Israel is this: You may be strong enough on your own to defend yourself, but as long as America exists, you will never, ever, be alone. Israel has an obligation to defend itself."

Reality check, Mr. Secretary of State! You do not set government policy! It is the Congress that sets government policies! Not you! Further, the Congress answers to the voters!

As I write this, the IDF forces are massing on the border with Gaza. The journalists are reporting that it is not a question of "if" the invasion of Gaza is going to take place, but "when". They are convinced that the "most brutal phase is about to begin". In this, they are so right!

History has shown that "urban warfare" is supremely brutal! Such invasions have consistently back fired! Narrow streets and piles of rubble, have proven to be to the advantage of the defenders! As well, high buildings provide a fine vantage point for the defence. Not to mention providing nests for snipers! The Nazis learned this lesson well, in Stalingrad!

Hamas has been goading the Israelis! They welcome the bombing! It creates the very pile of rubble that is so important to their defense! It also infuriates the civilian population! Countless civilians are now motivated, consumed with hatred, determined to defend their homes! To repel the Israeli invaders!

As Hamas has a huge quantity of missiles, some of them of fine quality, it is reasonable to assume that they also have other modern weapons. These could well include anti-tank mines, anti-tank guided missiles, and rocket propelled grenades, or RPG's. In urban warfare, in close quarters, such weapons can prove to be devastating!

As the invasion has proven, Hamas also has a great many drones. These drones have the virtue of being cheap and effective. They can be used for surveillance, and to drop explosives on people and machines.

No doubt, the Hamas fighters have been training, possibly for years, and are well aware of the terrain. They are also reported to have a most impressive network of tunnels.

The Americans can testify to the effectiveness of tunnels! In Viet Nam, the Vietnamese were able to hide deep underground, so that the American bombing was largely ineffective. Then, as the tunnels were extensive and elaborate, they were able to emerge at numerous places, to the shock and horror of their enemy! It is very likely that the Hamas fighters are safely hiding in their tunnels, while the Israelis bomb Gaza!

Hamas is likely also counting on receiving support from other Palestinians, as well as other countries in the area. The journalists are reporting that Israel also attacked military targets in Syria and Lebanon, as well as in Jerusalem and the West Bank! This is being referred to as a "major escalation of the war"! Israel is playing right into the hands of Hamas!

Strangely enough, this is not too surprising! As Lenin stated, in Imperialism, the Highest State of Capitalism, "imperialism is reaction, right down the line"! Israelis are monopoly capitalists, imperialists, and imperialists love war! They are also determined to drag as many other countries into the war, as they can! That includes America!

Yet the revolutionary movement is growing ever stronger. Now, more than ever, it is imperative that all working people, working class as well as middle class, get in touch with the politicians in Washington. Encourage the rank and file Members of Congress, to form an Independent Congressional Association. Remind them of their oath, to "preserve, protect and defend the Constitution"! They did not take an oath to protect the billionaires! It is the Party bosses, both Republican and Democratic, who are determined to serve the billionaires!

Remember your oath! Defend the Constitution! Do not allow Biden to go to war in the Middle East! That war does not concern Americans! Form an Independent Congressional Association!

CHAPTER 3

CONCERNING THE USE OF TERROR

Day Eleven of the Israel - Hamas War. The United Nations is describing the situation in the Gaza Strip as a "humanitarian disaster". The people within Gaza have been cut off from electricity, water, food and medical supplies. The morgues are filled to capacity. They cannot be buried fast enough. The dead bodies are now being placed in refrigerator vehicles.

A great many people have fled to Southern Gaza, in anticipation of an invasion by Israel. A great many others have not fled. Some of them are able bodied people, who have chosen to stay and defend their homes. Others do not have the choice of fleeing. These include the sick, crippled, disabled, elderly, pregnant women and people confined to hospitals. The medical personnel, those who are responsible for their patients, have chosen to stay and perform their duty. At the risk of their own lives, they refuse to abandon their patients.

The Israeli air strikes, in the Gaza Strip, are targeting Hamas, or at least the building that are thought to be used by Hamas. This has caused a great deal of destruction and death, mainly of civilians. As most of the Hamas fighters are safely underground, in bunkers, this bombing is largely ineffective.

This bombing campaign is similar to the policy carried out by the Americans, in their war in Vietnam, which spread to Southeast Asia. In that case, President Johnson stated his goal, to be that of "bombing the Vietnamese back to the stone age"!

Someone should have told Johnson, that the Vietnamese were not far removed from the stone age! They did not have far to go!

The goal of the American bombing campaign, was never clearly stated. Perhaps it was thought that by destroying their infrastructure, as well as by killing enough of their military personnel, the Vietnamese would embrace American rule, at least in South Vietnam. Or perhaps the American military simply did not think it through. They may have simply resorted to a bombing campaign *because they could!*

As is well known, that American bombing campaign backfired! It did not break their will to resist! On the contrary, this merely gave rise to hatred and bitterness, so that their determination to fight became ever stronger! Almost the entire population volunteered to assist, in any way possible. Countless men and women joined the military. Others worked on supplying the troops with everything they needed. They adapted. They improvised. They endured. They overcame.

The Vietnamese made great use of tunnels. These tunnels were extensive, deep and elaborate. The American artillery and air strikes against them, were completely ineffective. The use of "special forces", referred to as "tunnel rats", troops who were sent into those tunnels, proved to be a disaster. The Vietnamese were waiting for them. Very few of those men emerged alive, from those tunnels.

In anger and frustration, the American military decided to compound their mistake. As the Vietnamese were also making good use of neighbouring countries, the Americans decided to bomb those countries also! This had the effect of merely enlarging the war.

A similar situation is now happening in the Middle East. In addition to Hamas, which is active mainly in the Gaza Strip, it is thought that another resistance group, that of Hezbollah, is active in Lebanon. So now

Israel is launching artillery and air strikes into Lebanon. As well, parts of Syria and the the occupied West Bank, have also come under fire.

If it is the goal of the government of Israel, to enlarge the war, it is succeeding. The government of Iran is giving serious consideration to becoming involved!

The Iranian Foreign Minister has just given a clear warning. He is well aware that Israel plans to invade the Gaza Strip, and is considering a "pre-emptive action within the coming hours", on behalf of that which he refers to as the "Resistance Axis". Apparently, that is a reference to Iran, Syria, Lebanon, Hamas, Hezbollah and possibly other groups and countries, including Iraq.

In his statement, he made it clear that the "expansion of the war", into neighbouring countries, by Israel, would "change the situational geographical map of the occupation regime of Jerusalem".

He is also concerned with the involvement of others. "The Americans cannot constantly send the signal that they do not want the war to expand, but they are on the side of the Zionists".

It is significant that he refers to Israel as being a *Zionist* state, because that is precisely the case! A Zionist is someone who maintains that God gave the "Holy Land", now referred to as Palestine, to the Hebrews. Biblical scholars think this may have happened over three thousand years ago, perhaps during the reign of the Egyptian Pharaoh Akhenaten. This is of course a matter of faith, and all matters of faith should be respected.

Respect of faith is one thing, but should not be confused with facts. The fact is that, perhaps two thousand years ago, the state of Israel was destroyed, by the Romans. After that, Palestine was occupied by people of various faiths, including Jews, Muslims and Christians. By and large, they respected the beliefs of each other, and lived together in peace.

That changed dramatically, in 1948. In that year, certain members of the Jewish faith, those who are of the opinion that the Holy Land is for

Jews only, decided to form a country, and call it Israel. These people are referred to as "Zionists".

The Zionists took control of Palestine, declared it to be the state of Israel, and ordered all non Jewish residents to leave their ancestral homes. An estimated seven hundred thousand Palestinians were "displaced", to put it politely, driven from their homes. For the most part, they settled in places now referred to as the Gaza Strip and the West Bank.

Since that time, the state of Israel has expanded to include the West Bank, the Gaza Strip and the Golan Heights.

Of course, the people of Palestine have been resisting! This has given rise to various resistance groups, including Hamas, Hezbollah, and the Palestinian Liberation Organization. All of them are united in their opposition to the Zionists! They have no quarrel with those of the Jewish faith, or any other faith! Just the Zionists! It is important that all common people be made aware of this!

Without doubt, in the current Israel - Hamas War, both sides have resorted to acts of terror, in the form of attacks on civilians. It is correct to condemn such attacks on civilians! On the other hand, this is not to say that all acts of terror are to be condemned.

Lenin made this quite clear, in 1920, in a meeting with British workers. Lenin responded to their concern with that of "Red Terror" in Russia. In particular, those workers expressed surprise at the lack of "freedom of the press and of assembly", of the "persecution of Mensheviks and those who support them", within Russia.

The response of Lenin is instructive: "My reply was that the real cause of the terror is the British imperialists and their 'allies', who practiced and are still practicing a White terror in Finland and in Hungary, in India and in Ireland....Our Red Terror is a defence of the working class against the exploiters, the crushing of the resistance from the exploiters...Freedom of the press and assembly under bourgeois democracy is freedom for the wealthy to conspire against the working class, freedom for the capitalists to bribe and buy up the press".

This is to stress the fact that "Red Terror", against our *class enemies*, the billionaires, the imperialists, is correct and just! That is far different from the use of terror against civilians!

Yet the Iranian Foreign Minister is correct, when he states that the Americans want the war to expand! They are war mongers! They are supporting the Zionists! That is the reason they have sent two military fleets to the Mediterranean! They are determined to expand the war, to go to war with various countries in the Middle East, including Iran!

It is imperative that we oppose the American plans for expanding the war in the Middle East! The American government supports the Zionist government of Israel! They want to spread the war, across the Middle East! They must be stopped!

This is not to say that we support acts of terror, against working people!

The Israelis are Zionists, determined to crush the people of Palestine. The people of Palestine have every right to resist such an occupation! They also have every right to demand a return of their ancestral homes!

At the same time, we must also make it clear that the Zionists, those who are currently in charge of Israel, are not to be confused with the vast majority of Jews. Support for the people of Palestine should not be confused with anti semitism.

These facts should be made clear to all working people.

CHAPTER 4

REACTION IS PREPARING TO SEIZE POWER

Recently, on Remembrance Day, November 11, 2023, Donald Trump, the former president of the United States, gave a speech in the state of New Hampshire. That speech was nothing less than historic. For once, Trump spoke the truth. He revealed his plan to once again occupy the White House, and settle some old scores with his perceived enemies. Almost all members of the main stream press were shocked and horrified. With good reason, I might add.

This is the speech Trump gave:

"The threat from outside sources is far less sinister, dangerous and grave than the threat from within. We will root out the Communists, Marxists, fascists and the Radical Left Thugs that live like vermin within the confines of our country, that lie and steal and cheat on elections and will do everything possible, legally and illegally, to destroy America and to destroy the American dream. The real threat is not from the Radical Right, the real threat is from the Radical Left and it is growing every single day. The threat from outside forces is far less sinister, dangerous and grave than the threat from within. Our threat is from within."

One of the more astute, among the bourgeois journalists, had this to say:

"A mask off moment, for the disgraced, twice impeached, four times indicted former president. He is making his contempt for our democracy and its institutions and its people crystal clear, once again.

"An ex president who tried to overturn our democracy, now claims in front of supporters, that Americans here at home pose a greater threat to our country, than any of our adversaries abroad. He describes his political opponents as 'vermin'".

This journalist is correct when he states that Trump regards all of his "political opponents" as "vermin". Yet it is also a fact that he regards *anyone* who opposes him as vermin! That includes members of law enforcement, such as judges and prosecutors, and even their families!

This reference to "vermin", naturally leads to comparisons to other authoritarian leaders. Historians have this to say: "The language is the language that dictators use to install fear...When you dehumanize an opponent, you strip them of their constitutional right to participate securely in a democracy, because you are saying that they are not human. That is what dictators do".

Another historian said that calling people "vermin", was used effectively by Hitler and Mussolini, in order to encourage their followers to engage in violence.

In response to question by journalists, a spokesman for Trump said that the comparison to Hitler, was ridiculous. He then went on to say that "critics of Trump are sad, miserable, and their entire existence will be crushed, when Trump returns to the White House".

Even the spokesman for Trump admits that Trump plans to use the power of the office of presidency, in order to "crush his critics"! So how is that different from the actions of Hitler?

Those who were formerly very close to Trump, say that he is not the same man who became president, in 2017. At that time, he was completely unprepared to assume power. For that reason, certain members of his "inner circle" were able to exercise a certain measure of influence over him. Such is no longer the case.

Now, journalists are reporting that an "ultra right wing think tank", which they refer to as the "Heritage Foundation", is "grooming" Trump, preparing to set him up as the next president, through "Project 2025". The members of this "think tank" are referred to as the "handlers" of Trump. The handlers, in turn, think of Trump as their "asset", as is indeed the case. In fact, Trump is nothing other than a puppet, and those handlers are "pulling his strings".

Journalists report that possibly as many as 54,000 people have already been "prepared", to be placed in "various government agencies, including legal, judicial, defence, regulatory and domestic policy jobs". These people have been carefully screened, by the handlers of Trump. Those same handlers plan to purge anyone, within the government as well as outside, who is "viewed as hostile". They have that which is politely referred to as a "very unconventional and elastic views of presidential power and traditional rule of law".

The "unconventional" plan to seize power, involves having Trump invoke the Insurrection Act, on the first day of office. This will give him the authority to use the military against American citizens. The military will then build camps capable of housing millions of people. This would then be followed by sweeping raids, in which countless "undocumented people" would be "incarcerated" in those huge camps, in preparation for "mass deportations".

To help speed mass deportations, Trump is preparing an enormous expansion of a form of removal that does not require due process hearing. Part of that "preparation" involves scrapping the Constitution.

Further, it stands to reason that those who do not qualify for deportation, may deserve "special treatment". The Nazis found that "termination" met the requirement. It is very likely that Trump has a similar solution in mind, especially as anyone who opposes him is mere "vermin".

A former high ranking official in the Trump administration, one who served him loyally for more than a year, stated it quite clearly: "He lacks any shred of human decency, humility or caring. He is morally bankrupt, breathtakingly dishonest, lethally incompetent, and stunningly ignorant of virtually anything related to governing, history, geography, human events or world affairs. He is a traitor and malignancy in our nation, and represents a clear and present danger to our democracy and the rule of law."

It is significant that no high ranking elected politician, of the Republican Party, has dared to challenge Trump on these plans. Even the National Chairwoman of the Republican National Committee, has been intimidated into silence. As she answered, in response to a question, "I am not going to comment on the candidates and their campaign messaging".

As a journalist pointed out, even though it is her *duty* to respond to such questions, she prefers to "turn a blind eye" to a "would be dictator".

This defence of Trump, as he plans to seize power and set himself up as a fascist dictator, is not limited to the RNC. Even the members of his election campaign defend his "vermin comment", as well as his suggestion that the Chairman of the Joint Chiefs of Staff, of the American military, should be executed! After all, that Chairman did not go along with his plan for a military coup! As well, they even defend his plan to jail those who are investigating Trump! Both judges and prosecutors! Even though they are merely performing their duty!

In fact, Trump wants no part of "democracy"! He want to "terminate" the Constitution! He plans to set himself up as a fascist dictator!

For the benefit of the countless working people who are just now becoming politically active, I should mention that both Hitler and Mussolini were fascists. By definition, a fascist is defined as "a follower of a political philosophy characterized by authoritarian views and a strong central government- and no tolerance for opposing opinions."

As is quite well known, in the nineteen thirties, Mussolini formed the National Fascist Party of Italy, while in Germany, Hitler formed the National Socialist Workers Party, or Nazi Party. Not that the Nazis were Socialist or Workers, but that is the name they chose.

It is also well known that, in the twentieth century, Hitler made a speech, on January 30, 1939, in which he stated that "the outbreak of war would mean the end of European Jewry". Hitler was not joking. For once, he was telling the truth. In fact, Hitler clearly stated, in his book Mein Kampf, that which he planned to do.

Now in the twenty first century, we have another maniac, another fascist, making similar threats. Just as Hitler had a huge following, so too Trump has a huge following. Further, Trump is equally determined to carry through on his threats. It would be a mistake to under estimate that psychopath!

The main stream journalists are deeply alarmed with the rhetoric of Trump. They are well aware that he consistently refers to them as the "purveyors of fake news". As that is the case, Trump sees them as "members of the Radical Left", as "vermin" that are determined to "destroy America". For that reason, he is determined to "crush" them, once he returns to the White House. Trump is not joking!

This threat of fascism comes as no surprise, to any Marxist. After all, we live under a state of monopoly capitalism, which is technically referred to as imperialism. And, as Lenin stated, in his book Imperialism, the Highest Stage of Capitalism, "the political features of imperialism are reaction all along the line". Fascism is one form of reaction.

Lenin also stated that it is simply not possible to "reform" imperialism. It has to be destroyed.

It is reasonable to assume that the journalists, for the main stream press, are accurately reporting events which are happening. Most of them, but by no means all, are people of principle. They take great pride in their work.

As that is the case, then it is clear that the political situation has recently changed, quite dramatically. Previously, the ruling class of billionaires, the bourgeoisie, regarded Trump as a "nuisance", a "loose cannon". Now they have found a use for that "nuisance", They have decided to surround that "loose cannon" with a small army of handlers, and put him to work, in their service.

This is no doubt their idea of "killing two birds with one stone". On the one hand, their current method of rule, that of the Constitution and the "Two Party System," is not working. So they have to change their method of rule. On the other hand, "Trump the Nuisance", even though one of their own, has to be contained. So by corralling Trump and setting him up as a Dictator, a figure head President, while scrapping the Constitution, both problems are solved!

The fact that the billionaires can no longer rule in the old way, is an indication of the strength of the revolutionary motion. It has now reached well into the middle class. Now it is up to the middle class, along with the working class, the proletariat, to overthrow this completely reactionary class of billionaires, the bourgeoisie, and crush them, under the Dictatorship of the Proletariat.

In order to stress that importance, perhaps it is best to give the opinion of Lenin, as he stated in Left Wing Communism, An Infantile Disorder:

"The fundamental law of revolution, which has been confirmed by all revolutions and especially by all three Russian revolutions in the twentieth century, is as follows: for a revolution to take place it is not enough for the exploited and oppressed masses to realize the impossibility of living in the old way, and demand changes; for a revolution to take place it is essential that the exploiters should not be able to live and rule in the old way. It is only when the '*lower classes*' *do not want* to live in the old way and the 'upper classes' *cannot carry on in the old way* that the revolution can triumph. This truth can be expressed in other words: revolution is impossible without a nation-wide crisis (affecting both the exploited and the exploiters). It follows that, for a revolution to take place, it is essential, first, that a majority of the workers (or at least a majority of the class-conscious, thinking, and politically active workers), should fully realize that revolution is necessary, and that they should be prepared to die for it; second, that the ruling classes should be going through a governmental crisis, which draws even the most backward masses into politics (symptomatic of any genuine revolution is a rapid, tenfold and even hundredfold increase in the size of the working and oppressed masses -hitherto apathetic- who are capable of waging the political struggle), weakens the government, and makes it possible for the revolutionaries to rapidly overthrow it." (italics by Lenin)

Without doubt, the current situation satisfies the "fundamental law of revolution". The "upper class", in our case the billionaires, can no longer "carry on in the old way". They are being forced to change their method or rule. As well, the working class *and the middle class* now "realize the impossibility of living in the old way",

I include the middle class, for a reason. Donald Trump speaks for the billionaires, when he states that anyone who opposes him, is "vermin", to be "crushed". That includes most members of the middle class! Such people have every reason to be concerned! Trump means what he says! It is in the best interest of the middle class, to take part in the revolution!

For those who are reluctant to "take the plunge", feel free to face reality. The billionaires have decided to place Trump in the White House, scrap the Constitution, and declare martial law. Anyone who opposes them is to be classified as "vermin", and dealt with as such. Further, all banks but eight, and all businesses but five, are "Too Small To Succeed". All must collapse, their "assets" to be picked up by the billionaires. The middle class is about to be wiped out!

That begs the question: What role should the middle class play in the approaching revolution?

There are a great many things which should be done, so it is difficult to determine the most important. But as Lenin stated in What Is To Be Done?, "Every question 'runs in a vicious circle', because political life as a whole is an endless chain, consisting of an infinite number of links. The whole art of politics lies in finding and taking a firm a grip as we can of the link that is least likely to be struck from our hands, the one that is most important at the given moment, the one that most of all guarantees its possessor the possession of the whole chain".

In my opinion, the "key link" now, is to *raise* the level of awareness of the proletariat, so that at least the *most advanced workers embrace Scientific Socialism,* in the form of the *Dictatorship of the Proletariat!* There are several ways in which this can be done.

On the one hand, they must become class conscious, aware of themselves as a class, with class interests that are diametrically opposed to that of the billionaires. As the working class is quite highly cultured, a great deal can be accomplished through social media. Workers should be encouraged to read the Essential Works of Lenin, as well as the Communist Manifesto, by Marx and Engels.

Further, marches and demonstrations should carry banners and posters which call for revolution, Scientific Socialism and the Dictatorship of the Proletariat. Protests can target certain key locations, such as banks and government agencies, homes of government officials, for example. Entertainers can compose songs to be sung at those locations. Make the protests entertaining! The best way to educate people, is through entertainment!

Middle class lawyers, preferably experts on Constitutional law, can prepare the arguments, to be presented before the Supreme Court, that the 2020 federal election was Unconstitutional, as it did not follow the procedure laid out in the Twelfth Amendment to the Constitution. For that reason, Biden is a fraudulent President, and Harris is a fraudulent Vive President. Demand that the approaching 2024 federal election follow those procedures.

Those same legal experts can go on social media, and explain to the working people that the Twelfth Amendment makes no mention of any political party, of any popular vote, or of any District. Nor does it mention any "running mate". Further, the states have no right to meddle in a federal election.

Bear in mind that the main thing is to *raise the level of awareness* of the working class. The success, or failure, of the court case, is of secondary significance.

All Americans, working class as well as middle class, should be encouraged to join the two mainstream political parties, preferably as card carrying members. Such members determine the candidates for all political offices. Insist that the new card carrying members select a candidate of their choice, for all political offices. But do not challenge any member of The Squad, or Senator Sanders.

The goal here is not to flood Washington, or any state capital, with Leftist people, although that would be a bonus. The goal is to raise the level of awareness of the working people, to get them politically active. In the political struggle, those who still have faith in the parliamentary system, will soon learn that the billionaires are in charge, and fully intend to remain in charge!

The middle class people can also be supremely helpful in organizing Councils, also known as Soviets, as well as taking part in the creation of a true Marxist Communist Party, one which calls for the Dictatorship of the Proletariat. In both cases, a great deal of discretion is required.

At the time of the revolution, an insurrection will be required. This means that Councils are required to train a great many working people, before hand. Those with previous military or police training will prove to be most valuable. They can help train the "raw recruits". All must be not only trained in the use of various weapons, but also equipped. As I have gone into this in a previous article, there is no need to repeat it here. I will only add that Councils should also assist the working people, any way possible. In this way, they are sure to earn the respect and support of all common people.

The creation of a true Communist Party requires even more discretion. The billionaires, reactionaries one and all, are "taking to arms". They have decided to set up Trump as a dictator, so that citizens have no rights. That makes them more dangerous than ever. It also means the creation of a Communist Party, Dictatorship of the Proletariat, is even more urgent.

Middle class people have the computer skills to manage this, without personally coming together. They also have the capital to arm and train working people. Bear in mind that the billionaires have already decided to separate all others from their capital. Further bear in mind that any sacrifices made now, will not be forgotten. Middle class people have a bright future, but only under Scientific Socialism, the Dictatorship of the Proletariat.

CHAPTER 5

CONCERNING THE IMPORTANCE OF A TRUE COMMUNIST PARTY, DICTATORSHIP OF THE PROLETARIAT

Of late, the mainstream press - the bourgeois press - has been howling that Trump is a "threat to democracy". This expression has even become so common, we can honestly say that it can be referred to as a "slogan".

Most common people - working people - may well get the idea that Trump, and only Trump, is a threat to their democratic way of life. The bourgeois journalists certainly go to considerable length, to give that impression. Now they are further howling that Trump is a fascist, and plans to set himself up as a dictator. There is some truth to all of this, as well as a great deal of distortion. It is the purpose of this article to clarify the situation.

Incidentally, for the purposes of this article, I refer to all working people, proletarians and farmers, as common people, as that is the manner in which they refer to themselves. I should also mention that the journalists who work for the mainstream press, are in the service of the capitalists, the billionaires, those who own that press. For that reason, they have to "slant" the news in favour of their bosses, regardless of their personal sentiments. For our part, it is important to separate the facts they state, from their analysis, which always supports the capitalists.

Especially now, it is vitally important to raise the level of awareness of the working class, the proletariat. As a result of the revolutionary motion, countless common people, those who were formerly apathetic, are now politically active, or as they phrase it, they "woke up". They are demanding change! They want answers! The only "answers" they get from the bourgeois press, are evasions and half truths.

It is of *vital importance* to *raise their level of awareness!* The reason I say this is because, as Lenin stated, "the only basis that can guarantee our victory…is the *socialist consciousness of the working class*". (my italics).

In previous articles, I documented the fact that "democracy" is merely a method of class rule. In fact, the "democratic republic" under which we live, is nothing other than the rule of the capitalists, the billionaires, over the common people, the working class. This particular method of rule has served them very well, for a great many years.

Yet, that "time honoured" method of rule, is no longer effective. The revolutionary motion has inspired countless working people, those who were formerly apathetic, to become politically active. The two party system is no longer working. In Washington, members of the two parties are "behaving like third graders", as the journalists phrase it. As one politician stated, rather politely, the Members of Congress are "demonstrating complete disfunction, that has descended into physical violence".

This is not an exaggeration. In fact, one House Member physically elbowed another! One Senator even challenged the leader of a trade union to "fisticuffs"! The comparison to children is a school yard is legitimate!

So now the ruling class of billionaires, the bourgeoisie, has decided to scrap their current method of rule, that of the democratic republic, complete with a Two Party System, in favour of a fascist dictatorship, with Trump as the figurehead leader.

It is important to remember that Trump does not write his own speeches. Nor does he speak "off the cuff", so to speak. On the contrary, Trump reads from a teleprompter, one that the television cameras are careful to conceal. As a highly successful entertainer, Trump is able to read from that teleprompter, without appearing to do so. His years in show business are serving him well! It is only after the "show is over", that Trump posts on social media, that which he has recited. In this way, common people get the idea that Trump writes his own speeches.

The point to be stressed, is that Trump is merely the "figurehead", the "puppet", for the billionaires. The class conflict has not changed! It remains class warfare! It remains a conflict of the working class, the proletariat, against the billionaires, the bourgeoisie! It is *not* a conflict between the working class and a "wannabe" dictator!

It is also a fact that the current revolutionary movement is nothing other than "class consciousness in embryonic form". This movement testifies to the "awakening antagonism" between the working class, and the capitalist class of billionaires. This is *not* to say that the working class is now "class conscious". It most certainly is not!

In fact, class consciousness can only be *brought* to the working class, from an *outside source*. The working class, exclusively by its own efforts, is able to develop only trade union consciousness. Working people may recognize the fact that there is strength in numbers, hence the importance of trade unions.

By contrast, the theory of scientific socialism, including the Dictatorship of the Proletariat, was elaborated by Marx and Engels. These men were well educated, intellectual members of the "propertied classes". It is to their credit that they turned their backs on a life of comfort, as members of the middle class, to instead work in the service of the common people.

These revolutionary theories, that of scientific socialism, are well known to the "upper classes", the billionaires, as well as the upper middle class.

After all, those theories are taught in university. They are certainly not taught, or even mentioned, anywhere else! And as very few working class people can afford to attend university, members of the working class are largely unaware! Fine by the billionaires!

Now the billionaires have thoughtfully simplified the class struggle, with their plan to wipe out the middle class. Their plan is to ruin thousands of banks, as well as tens of thousands of businesses. That includes General Motors, as GM is one of the businesses that is *Too Small To Succeed!* As they fail, the billionaires will "pick up their assets". In the process, the middle class will be ruined, forced into the ranks of the proletariat. Anyone who objects to this, will be classified as "vermin", and will be treated as such.

If nothing else, this provides the intellectual members of the middle class with the motivation to take part in the creation of a *true* Communist Party, Dictatorship of the Proletariat, CP,DP. (The only true Communist Party is one which calls for the Dictatorship of the Proletariat)

The current plan of the billionaires, is to create a fascist state, with Trump as the figure head dictator, abolish the Constitution, and wipe out the middle class. They must be stopped. That calls for a truly fine organization, in order to establish Scientific Socialism. The mass revolutionary movement does not *relieve* us of this burden. On the contrary, it "*imposes*" this duty on us, "because the spontaneous movement of the proletariat will not become a genuine 'class struggle' until it is led by a strong organization of revolutionaries". (Italics by Lenin)

Without a true revolutionary organization, the *spontaneous movement* of the working class, towards socialism, is certain to be diverted, by the bourgeois ideology, onto some harmless path of social reform. After all, the "bourgeois ideology is far older in origin than Social Democratic (Marxist) ideology; because it is more fully developed and because it possesses *immeasurably* more opportunities for being distributed". (Italics by Lenin)

Having said that, perhaps a word of caution is in order. Taking part in the creation of such a truly revolutionary organization, a true Communist

Party, is not something to be taken lightly. Once that step is taken, *there can be no turning back!*

In the case of Russia, under the rule of the Tzar, Lenin gave five requirements, which are necessary for a true Marxist Communist Party. In the case of twenty first century America, such a similar situation may also soon exist, *unless the billionaires are stopped!* Here are those requirements:

1) "That no movement can be durable without a stable organization of leaders to maintain continuity

2) "That the more widely the masses are spontaneously drawn into the struggle and form the basis of the movement and participate in it, the more necessary is it to have such an organization, and the more stable must it be (for it is much easier for demagogues to side track more backward sections of the masses)

3) "That the organization must consist chiefly of persons engaged in revolutionary activities as a profession

4) "That in a country with an autocratic government, the more we *restrict* the membership of this organization to persons who are engaged in revolutionary activities as a profession, and who have been professionally trained in the art of combatting the political police, the more difficult will it be to catch the organization and

5) "The *wider* will be the circle of men and women of the working class or of other classes of society able to join the movement and perform active work in it" (italics by Lenin)

Those who take part in creating, and becoming a Member of, a true Communist Party, must be devoted revolutionaries. This is not to say that such people cannot also be members of other professions. In fact, this is recommended. After all, everyone has to earn a living. Bear in mind that Lenin was also an attorney.

As yet, we still live under a democratic republic, complete with some restricted democratic rights, as guaranteed in the Constitution. That democratic republic is under an even greater threat than I suspected, according to the latest internet report.

Perhaps I should add that I do not generally go into any great detail, concerning my sources, as I consider such details to be secondary. Besides, I do not want to face a law suit.

A respected journalist is reporting that, as part of "Project 2025", also known as "Agenda 47", no less than eighty people are busy "screening" candidates, for positions in the "soon to be created" Trump administration. One of these busy souls is a man who was formerly a lawyer for Trump, now in serious legal trouble. No doubt, he anticipates that with Trump once again in the White House, and the Constitution a "dim and distant memory", he will not have to worry about any pesky legal charges.

The purpose of this screening, of thousands of people, is to determine those who are truly "devoted and loyal", worthy to serve in a Trump administration. As one source admitted to reporters, "Hundreds of people are spending millions of dollars to install a pre vetted pro Trump army of up to 54,000 loyalists, across government, to rip off the restraints imposed on the previous 46 presidents. The screening for ready to serve loyalists has already begun, driven in part by artificial Intelligence from a (technical company), contracted for the project. Social media histories are already being plumbed. Trump himself spends little time plotting government plans, but he is well aware of a highly coordinated campaign to be ready to jam government offices with loyalists willing to stretch traditional boundaries".

The expression, to "stretch traditional boundaries", is a reference to the plan, by the billionaires, to install Trump as a figurehead president. This loud mouthed fool can then be their puppet, with the billionaires pulling his strings. They further plan to scrap the Constitution, and set Trump up as President For Life, a Dictator.

We have always needed a true Communist Party. That is certainly true today, more so than ever. In fact, it is urgently needed, as it is far more difficult to establish such a Party, under a dictator. The billionaires plan to set up such a dictator, "at or before" the 2024 election, according to Trump. Feel free to take him at his word! He means it! That man does not have a sense of humour!

In terms of forming a Communist Party, Dictatorship of the Proletariat, the "ball is in the court" of the middle class, if you will excuse the sports metaphor. It is highly unlikely that any members of the working class could manage that, although certain advanced workers can no doubt be helpful. Almost all proletarians simply do not have the training necessary, to accomplish such a task.

Yet the revolutionary motion continues to grow, becoming ever stronger. The country has been referred to as a "powder keg", in that any "spark" can cause an explosion. It is just a matter of time before a full blown revolution breaks out. The success - or failure!- of that revolution, depends largely upon the willingness of intellectual members of the middle class, to take part in creating a true Communist Party, Dictatorship of the Proletariat, CP, DP.

The decision, of middle class intellectuals, to establish a true CP,DP, should not be terribly difficult. After all, the billionaires have already decided to set up a dictator, and wipe out the middle class, by separating them from their capital. This will have the effect of driving all middle class people into bankruptcy, and the ranks of the proletariat.

By contrast, a true CP,DP can properly lead the forth coming revolution. For that reason, it is almost certain to succeed. Middle class intellectuals who support the revolution, whether Communists or not, will be rewarded, under the Dictatorship of the Proletariat. They will be placed in key management positions, running various businesses.

This revolution will be thorough. Bourgeois intellectuals will be a target of that revolution. That includes bourgeois scholars and bourgeois scientists. Their spiritual power, which I refer to as "professor power", will be broken. The act of breaking that spiritual power, will not be pleasant.

Middle class intellectuals who oppose the revolution, will be considered to be supporters of the billionaires, enemies of the people, and treated accordingly.

To all middle class intellectuals, I have a word of advice: Choose Wisely!

CHAPTER 6

CONCERNING THE "DISFUNCTION" WITHIN THE REPUBLICAN PARTY

Allow me to start this article, by first expressing my most heartfelt gratitude to the capitalists, the billionaires, for inventing the internet. It is truly a most revolutionary invention. Perhaps the best way to express my appreciation, on behalf of the working class, is by using the internet to expose the lies and hypocrisy, of those same billionaires.

The journalists who have websites, on the internet, are not constrained by the same rules, under which mainstream journalists labour. The unfortunates who work for the mainstream media, are forced to bias the news, in favour of the capitalists. After all, it is the billionaires who own those international media outlets. Either report the news the way the capitalists want it reported, or look for a job elsewhere!

For that reason, it is only the "freelance" journalists who report on subjects which the capitalists would prefer to keep quiet. One of these topics is that of the "disfunction" within the Republican Party, the Grand Old Party, the GOP.

One such report I found to be most interesting, for several reasons.

The journalist starts by stating that "Republicans are getting crushed at the state level, after letting MAGA take over". Of course, MAGA stands for Make America Great Again, the "war cry", or slogan, if you prefer, of Trump and his supporters. He goes on to give several examples. This includes the state of Arizona, in which the state GOP is "begging the national GOP for money". Then, in the state of Michigan, he reports that "a brawl broke out, an actual physical fight". As well, in the state of Georgia, "extremism issues are leading to similar electoral failure". His conclusion, is that "donors want nothing to do with those people". He then quite gleefully pointed out that the Democratic Party is raising over double the money, as the Republican Party.

It is reasonable to assume that this fellow considers himself to be "Leftist", or at least "Left leaning", possibly even socialist. He appears to "rejoice" over the fact that "donors want nothing to do with those people", meaning the MAGA people, so that it works in favour of the Democrats. This outlook is a faulty.

As is well known, there are two mainstream political parties, Democratic and Republican. The Democrats would have us believe that they represent the "middle class", so that they are the Party of the "Left". The Republicans would have us believe that they represent the "business" people, so that they are the Party of the "Right".

It is significant that the term "middle class" is referred to rather vaguely, as it is also customary to deny the existence of classes, here in North America. As well, the term "business" is equally vague. Both terms are used to spread confusion.

In fact, both Parties serve the same class, that of the monopoly capitalists, the billionaires, the bourgeoisie. Tweedle Dee and Tweedle Dum. The one and only difference between them, is in the spelling of the words.

Politicians, for both Parties, are "managed", by a small army of professional people, referred to as "handlers". These include writers, who prepare their speeches. Others are responsible for their hair, clothes and makeup. Still others tend to the proper posture, tone of voice, facial expression and hand gestures of their "assets". Politicians who have previously pursued successful careers, in the field of entertainment, tend to make the best "assets". It is no coincidence that Trump is a highly successful entertainer!

As for the money received from "donors", there are "strings attached". Those who donate that money, expect something in return! These donors are invariably capitalists, and see these "donations" as an investment. The particular Party which delivers a return, upon that investment, is a matter of no concern, to the capitalist.

There can be no doubt that all elections are expensive. Candidates who are elected, owe a great deal to their donors! By far, the most expensive election, is that of the federal. The federal election of 2020 cost a record breaking *$14.4 billion!*

With that in mind, and in the interest of raising the level of awareness of the working class, especially those who still have faith in the democratic process, allow me to make a few suggestions.

Instead of condemning the "MAGA" people, for joining the Republican Party, feel free to follow their lead! Encourage all working people, especially all those that are "Leftist", to join the two mainstream political parties, both Democratic and Republican, preferably as card carrying members. It is the card carrying members of each Party, who decide the candidates for any and all political offices. If enough Leftist people join both Parties, we can make sure that Leftist candidates are on the ballot, for both Parties. In that manner, it will not matter how much money was donated, to either Party. A Leftist candidate is certain to win!

At the same time, feel free to associate with other members of those parties, including those working people who are considered to be MAGA. Let them know that we are not the demons that their leaders make us out to be!

Bear in mind that it is "against the rules" to become members of both Parties, but not against the law. Feel free to be naughty!

Another suggestion is to encourage middle class attorneys, experts on Constitutional law, those who consider themselves to be "Leftist", to challenge the 2020 federal election, in the Supreme Court. I have previously made suggestions, on the grounds that it did not follow the procedure laid out in the Twelfth Amendment to the Constitution. Demand that the next federal election follow those procedures. At the same time, flood social media with updates, as the mainstream press is certain to give their own slant to the story.

If the Supreme Court rules in our favour, then any popular vote will be declared to be meaningless, and the states will not be allowed to meddle in any federal election. The electors, appointed by each state, will be allowed to vote for the candidates for President, as well as Vice President, of their choice. Such candidates may, or may not, be members of any political party.

As I have given a list of possible arguments, to be used by these attorneys, there is no need to repeat it here. Besides, those people will, no doubt, choose their own arguments.

Bear in mind that the vast majority of people are working class, proletarians. As Lenin pointed out, "only the socialist consciousness of the working class can guarantee our victory". This is to stress the fact that *not* all MAGA people are the enemy! By no means! There are a great many supporters of Trump, MAGA people, common people, who are merely misled. They tend to be less advanced, and require patience. More popular literature must be prepared, with them in mind, and possibly sent through email, if appropriate.

In conclusion, it is up to all Leftist people to take advantage of the chaos and confusion, within both political parties. It is a reflection of the chaos taking place, within the ranks of the billionaires. Use this as a means to raise the level of awareness of the working class.

The ruling class of monopoly capitalists, the billionaires, the bourgeoisie, can no longer rule in the old way. They are now determined to change their method of rule. That new method of rule involves scrapping the Constitution, and setting up Trump, as a fascist dictator.

The key to stopping them, is to raise the level of awareness, of the working class, the proletariat, to that of Scientific Socialism. Let us focus our every action, with the idea of raising their level of consciousness. If those of us who are true Scientific Socialists, Communists, do our job properly, then all working people will soon be discussing the Dictatorship of the Proletariat.

CHAPTER 7

WAR IN THE MIDDLE EAST

Since the start of the Israel-Hamas war, the press has been focused on the possibility that this may be the beginning of World War 3. There is cause for their concern, as the war is clearly expanding. The American military sent two fleets to the Mediterranean, in preparation for a wide ranging war.

According to a statement by the American Defence Secretary, President Biden ordered two airstrikes in Syria, "against Iran and its aligned groups". As he phrased it, "The president has no higher priority than the safety of U.S. personnel, and he directed today's action to make clear that the United States will defend itself, its personnel, and its interests."

The press went on to explain that these airstrikes are an attempt to "quell wave after wave of drone and rocket attacks against American troops, based in Syria and Iraq, triggered by the Israel-Hamas war." They further report that there are 900 American troops in Syria, and 2,500 more in Iraq. Their mission is to "advise and assist local forces trying to prevent a resurgence of an 'Islamic State', which in 2014 seized large swaths of both countries, but was later defeated".

As well, Israel has carried out "strikes" against a military organization referred to as Hezbollah, located to the north of Israel, in the southern part of the country of Lebanon. The press refers to these strikes, as a response to "tense exchanges of gun fire, along the Lebanon border". In addition, Israel has also prepared "elite forces", in anticipation of invading Lebanon.

As it is also accepted that Hezbollah is the dominant political force in the West Bank, which is currently occupied by Israel, there are fears that Israel could also be attacked, from the West Bank.

In Yemen, the Houthis have declared themselves to be part of an "Axis of Resistance", of Iranian allies, determined to "retaliate on the Israeli war on Gaza".

The Iranian Foreign Minister just released a statement, to the effect that, "Resistance leaders in the region say that until the full rights of the state of Palestine are realized, and until an outcome is reached in confronting the occupation in the region, they will keep their fingers on the trigger".

The press is also reporting, that the Foreign Minister of Iraq issued another warning, threatening the expansion of the war, if the current cease fire does not continue.

As I write this, both Israel and Hamas have agreed to a "tentative truce", for several days, with the goal of exchanging hostages, and providing the people of Gaza with desperately needed humanitarian supplies.

On the first day of that truce, a number of hostages were exchanged, and a number of trucks were allowed to enter Gaza, from the border with Egypt. It remains to be seen if this truce will continue to hold. If the truce fails to hold, then Iran is threatening an escalation of the war.

No wonder the press is so concerned with the threat of World War 3!

As that is the case, perhaps it is best to examine the events which led up to the two previous world wars. If those events are similar to the current events, then we have good reason to worry.

No doubt, all middle class intellectuals will find this to be supremely boring, as they are already well aware. Yet my main concern is with the recently awakened members of the working class, as they have to be "brought up to speed", so to speak. Or as Lenin stated, "our very first and most imperative duty is to help to train working class revolutionaries who will be on the same level *in regard to Party activity* as intellectual revolutionaries".(italics by Lenin) After all, in order for the approaching revolution to be successful, a truly *Scientific Socialist Revolution,* the proletariat, or at least the most advanced strata of the proletariat, must be class conscious.

It is safe to say that the events leading up to the First World War, started in the nineteenth century. In fact, Marx and Engels lived and worked in the mid to late nineteenth century, during a time in which capitalism was in its early, competitive stage. At that time, the capitalists of the so called "Great Powers", the most highly industrialized countries of the world, were busy "dividing up the world", between them.

As a result of their study of capitalism, Marx and Engels wrote the Communist Manifesto, in 1848. They documented the fact that the industrial revolution gave rise to capitalism, and further, that capitalism initially played a most revolutionary role. In fact, at first, it had *certain* progressive characteristics. This is explained quite clearly, in their Manifesto.

Just as Marx predicted, capitalism developed into the stage of monopoly, around the "turn of the century", which is to say the beginning of the twentieth century. At that time, the most successful capitalists, those who were then referred to as "multi millionaires", agreed to cooperate, as it was thought that this was the best way to ensure a huge profit, through "price fixing".

Those same monopoly capitalists were also aware of the fact that "something had changed", although they could not say exactly what that was. Not that they had any great desire to do so. They were only concerned with their profit.

Yet this "new state of affairs", under monopoly capitalism, became known as "imperialism". Those who choose to embrace monopoly capitalism, are referred to as "imperialists".

It was Lenin who conducted a thorough, scientific examination of monopoly capitalism, that of imperialism. This gave rise to a true masterpiece, titled Imperialism, the Highest Stage of Capitalism. Among other things, he concluded that imperialism is "reaction, right down the line".

At the same time that capitalism was reaching its highest monopoly stage, the "Great Powers", succeeded in "dividing up" the whole world, between them. All other countries became mere colonies of those highly industrialized, imperialist countries.

This led to a little problem, among the imperialist powers. As there were no more colonies available to be seized, this just meant that the imperialist powers that were "ambitious", anxious to "extend their influence" through out the world, had to separate certain colonies from other countries, other imperialists.

In particular, Germany was anxious to expand. With that in mind, a "Quadruple Alliance", otherwise known as the "Central Powers", took shape. These consisted of the countries of Germany, Austria-Hungary, the Ottoman Empire and Bulgaria.

In opposition to the Central Powers, the "Allied Powers", or "Entente Powers", came together, in the form of France, Russia, the United Kingdom, the United States, Italy and Japan. After all, these countries were not at all anxious to part company with their colonies!

This attempt to "redivide the world", among the "Great Powers", gave rise to the First World War, the "Great War", the "War To End All Wars". It did nothing of the sort! Yet, starting in 1914, millions of common people, workers and farmers- peasants- were engaged in a great slaughter! They were sent to kill each other, so that one or another group of monopoly capitalists, imperialists, could rule the world! The "Great War" was nothing other than the "Great Slaughter" of common people!

Yet it did give rise to revolution! In Russia, the Romanov Dynasty was first overthrown! At that point, the Russian capitalists seized power! They were determined to continue the Great Slaughter of common people! In the interests of seizing colonies and increasing profits, of course!

In response, the workers and poor peasants rose up, under Lenin, overthrew the Russian capitalists, and created the first Scientific Socialist republic in the world! As the vast majority of common people were poor peasants, this became known as the Dictatorship of the Proletariat and Poor Peasants! Thus was born the Union of Soviet Socialist Republics, led by Lenin!

After the death of Lenin, Stalin took over the reigns of power, and continued to crush the capitalists. He carried out the policies started by Lenin.

We can only stress the fact that the Soviet Union, under Lenin and Stalin, was a truly Scientific Socialist country, a Communist country, one which embraced the Dictatorship of the Proletariat. It was only after the death of Stalin, in 1953, that the Russian capitalists were able to restore capitalism, and place the fool Khrushchev in charge.

Events which happened, immediately after the first successful socialist revolution in Russia, are instructive. These serve to drive home the fact that, after any socialist revolution, the monopoly capitalists, the billionaires, do not "resign themselves to their fate". On the contrary, there "resistance increases tenfold", as they make every effort to "restore their paradise lost", according to Lenin.

In fact, Lenin was shot, in 1918, and as a result of complications from those wounds, he died in 1924. Also, in Germany, in 1919, Karl Liebnecht and Rosa Luxembourg were murdered. Other working class revolutionary leaders, around the world, were also murdered. Still others, such as Karl Kautsky, "turned his coat", becoming a traitor to the proletariat. He and numerous others "sold out"! As I have covered this is previous articles, there is no need to repeat it here.

The point is that the working class was deprived of some of their finest leaders! As a result of this, the international working class movement, towards Scientific Socialism- temporarily!- regressed.

At that time, immediately after the Great Russian October Socialist Revolution, the most advanced strata of the proletariat, had embraced the principles of Scientific Socialism! The Dictatorship of the Proletariat was widely discussed! Such is no longer the case!

This is no cause for despair on our part! The working class, at least in the most highly industrialized countries of the world, are quite highly cultured. Most are literate, and have access to digital devices. The internet is "at their finger tips"! That makes the task of raising their level of awareness, to the level of Scientific Socialists, ever so much easier!

This brings us to the subject of the Second World War, starting in 1939. Once again, the imperialist powers were "at each others throats", squabbling over the colonies. For that reason, the slogan of the true Communist Parties, of all the warring countries, was "Turn the Imperialist War Into A Civil War". By a civil war was meant a revolutionary war, of Scientific Socialism, against the ruling class of capitalists, as had happened in Russia, in 1917. This was correct, at the start of the war, with Germany and Italy, waging war on France and Great Britain.

That changed dramatically on June 22, 1941. On that day, Nazi Germany invaded the first Scientific Socialist Republic, the Soviet Union. The imperialist world war was converted into a just war! The Socialist Soviet Union had to be defended! The slogans, of the true Communist Parties, of all countries, changed to "Defend the Socialist Soviet Union!"

In response to the invasion, the Soviet Union called for all citizens to defend the socialist homeland! They referred to this as the Great Patriotic War! It was no longer a war to redivide the world! The war which started, in 1939, between the imperialist powers, was converted into a just war! The Socialist Soviet Union had to be defended!

Of course, as is well known, the Soviet Union defeated the Nazi invaders, in 1945. The alliance, between the Communists of all countries, and the imperialists of the countries at war with Nazi Germany, came to an end.

It is also well known that Stalin died, in 1953. At that time, the Russian capitalists were able to restore capitalism, within the Soviet Union, due to mistakes that Stalin had made.

The Chinese Communists conducted a proper criticism of Stalin, and concluded that he was a fine revolutionary, but had made a number of serious mistakes, which allowed the capitalists to return to power, after he died.

After Mao died, in 1976, the Chinese capitalists were also able to return to power, due to mistakes that Mao had made. As I have documented this in a previous article, there is no need to repeat it here.

Suffice it to say that the road to Scientific Socialism, is not a paved highway. Our class enemies, the monopoly capitalists, the billionaires, are not about to "resign themselves to their fate", after the revolution! Hence the need for the Dictatorship of the Proletariat! The fact that they were able to return to power, in both the Soviet Union and China, leaves no room for any doubt!

It is up to us to honour the memory of previous great revolutionaries, Stalin as well as Mao, by learning from their mistakes!

Of course, those who restored capitalism in the USSR and in China, do not refer to themselves as capitalists. They call themselves "Communists". As capitalism is failing in both of those countries, the western capitalists offer this as "proof" that Communism is a failure!

Nothing could be further from the truth! It just means that capitalism is failing in the countries which were -formerly!- socialist!

I can only hope that our recent converts to Scientific Socialism, find that to be helpful.

Now to return to the subject of the Middle Eastern War, and the threat of World War 3.

The fact that the American imperialists have troops located in both Syria and Iraq, is an indication that they consider the Middle East to be part of their "sphere of influence", to use their stilted expression. It is also clear, from their press reports, that the Americans consider Iran to be the main threat to that "influence"!

For the moment, there appears to be no other imperialist power, which they refer to as a "super power", competing to take control of the region. Yet no doubt, numerous countries, in the Middle East, are fighting for the right to self determination. They want to be free from the influence of the American imperialists.

As well, the people of Palestine are fighting to throw off the yoke of the Zionists. They want to create a homeland of Palestine, one in which people of various religions can live together, in peace. This includes the Muslims, Jews and Christians. They previously lived together, for hundreds of years, respecting the beliefs of each other, and plan to do so again. But then the Zionists have to be first overthrown!

Both Israel and Hamas are making the mistake of resorting to the use of terror, against civilians, in the current war. Such a use of terror cannot be justified. For that reason, the leaders of both factions could soon be facing charges, from the International Criminal Court.

In summary, we can conclude that the war has already spread to other countries in the Middle East, most notably Lebanon and Syria. It will very likely spread to more countries, mainly in opposition to the Zionists and American imperialism. That is a far cry from World War 3.

Just how far the war spreads, is largely up to the revolutionary forces in America. If Trump manages to return to power and establish his fascist dictatorship, then it is very likely that the current Middle Eastern War will expand, as the world unites against American Fascist Imperialism.

Yet the revolutionary forces within America are determined, that will not happen!

CHAPTER 8

GROWING ANTI FASCIST MOVEMENT IN AMERICA

There is growing concern, in America, about the MAGA, Make America Great Again, movement. Even one of the most highly respected newspapers in the country, expressed fear that the country was threatened with the "end of democracy". These fears are well grounded!

MAGA was the slogan, which was coined by the handlers of Trump, on his first successful run for the presidency. It has a certain appeal, at least to the less advanced, more patriotic Americans, as it implies that America was great, at one time, but has since fallen into decline. It also implies that Trump is able to restore that "greatness".

As the vast majority of working people are struggling to "make ends meet", this promise, by a man with great personal charm, has a certain appeal. In the past, in other countries, under similar circumstances, similar appeals, by similar leaders, also of great personal charm, led to fascist dictatorships. The concern is that America is also heading towards a fascist dictatorship.

The journalists are documenting the opinion of historians, that "Trump's rise to power, was almost immediately accompanied by debates over whether his ascendency, and the ascendency of other leaders around the world, with similar political views, signalled a revival of fascism. Fascism is generally understood as an authoritarian, far right system of government in which hyper nationalism is a central component. It also often features a cult of personality around a strongman leader, the justification of violence or retribution against opponents, and the repeated denigration of the rule of law".

It is no secret that, as soon as Trump is once again President, he plans to immediately declare martial law, and to scrap the Constitution. In fact, he is determined to do this, on his first day back in the White House. This scrapping of the Constitution is more politely referred to as the "denigration of the rule of law".

It is safe to say that this "cult of personality", is also an accurate description of the followers of Trump. We can further say that, just as "past fascist leaders have appealed to a sense of victimhood, to justify their actions", so too Trump is "following in the footsteps of previous fascists". Trump thinks that no one has any right to charge him with any crime! He thinks he is above the law! He fully intends to "get even" with those who are pressing charges against him! That includes the prosecutors and judges!

The historical experts are agreed that fascism contains certain elements, that make it "far more dangerous than authoritarianism". One of those elements is "rejection of democracy, in favour of a strongman".

Of course, that "strongman" is Donald Trump, whom the billionaires plan to send to the White House, in order to "maintain stability", or "law and order", as they phrase it. In the process, the Constitution is to be scrapped, Trump is to be empowered as an absolute dictator, with unlimited powers, while "social order" is to be maintained, through the use of force.

The journalists are correct, when they state that, "for the fascist, war and violence are a means of strengthening society, by culling the weak and glorifying heroic warriors".

Perhaps one of the strangest features of fascism, is that of fostering resentment and anger against the "cultural elites", for causing so much pain and suffering, among the common people. This is encouraged, by those same "cultural elites"! Trump is one of those people, a member of the same class of "cultural elites", a billionaires, that he is raging against!

A strict hierarchy is also characteristic of fascism, based around male dominance, with women relegated to subservient roles. The men are "protectors, providers and controllers of the family". Any "deviants", such as homosexuals, are to be "culled". A "superior race or ethnicity", is glorified. In the case of America, under Trump, that ethnicity has fallen to "White Christian Nationalists".

In order for a fascist dictatorship to thrive, it is first necessary to normalize violence, denigrate all opponents, portray them as less than human, or "vermin", as Trump referred to anyone who opposes him. Then it is a simple matter of obliterating all opposition, through violence, as this is "justified", on the grounds that they are "vermin".

This turn to fascism, on the part of the ruling class of billionaires, the bourgeoisie, is in response to the fact that they can no longer rule, in the old way. So fascism is the alternative method or rule, which they have chosen. Trump is merely a figure head, a "puppet", with the billionaires "pulling the strings".

Yet a number of people and organizations, have come out in opposition to this. The most vocal are also the most "Right Wing", or "Conservative", as they describe themselves. Yet as they have taken a progressive position, it is up to "Leftist" people, to support them, in their opposition to this fascist trend.

No doubt, there will be a great howl from some of those on the "Left", to the effect that we cannot possibly work with such people. This stand is incorrect. As Lenin stated, "Only those who have no self reliance can fear to enter into temporary alliances even with unreliable people; not a single political party could exist without entering into such alliances."

Equally without doubt, such self described "conservatives" certainly qualify as "unreliable people". Yet these same conservatives are fighting against fascism, for our democratic rights. But, as Lenin also stated, "an essential condition for such an alliance must be complete freedom for Socialists to reveal to the working class that its interests are diametrically opposed to the interests of the bourgeoisie".

The following is a partial list of organizations which are opposing Trump, and his fascist policies. According to the internet, I have chosen to give a summary of their goals, as per their web sites:

"Our Principle Pac was a super pac established in January 2016 as part of the Never Trump Movement"

The Club For Growth is a self described "leading conservative organization, focused on American economic issues, limited government and conservative public policies".

"Americans Against Trump -We are the first people in the country doing the most to keep Donald Trump away from the presidency or kick him out of the White House. Join the resistance against the worst president in American history."

"Never Trump Movement is an ongoing conservative moderate movement that opposes Trump, started by Never Trump Republicans, and offer prominent Republican conservatives as GOP nomination for president"

The organization with the most detailed website is referred to as "The Lincoln Project", those who are clearly in opposition to Trump, and his fascist policies. I have chosen to copy the whole website:

"WE HAVE SEEN THE BEST AND WORST OF AMERICAN POLITICS

"The Lincoln Project is dedicated to the preservation, protection and defence of democracy. Since our launch in December, 2019, The Lincoln Project has played a unique role in American politics. We entered the political arena with two stated objectives: first, to defeat Donald Trump at the ballot box in 2020 (Done) The second was to ensure Trumpism failed alongside him. We understand that the real fight against Trumpism is just beginning and we're committed to seeing it through. Our democracy depends on it.

"Founded by former Republican strategists who understood the grave threat of Trumpism to our nation, the Lincoln Project team has expanded to include individuals from across the political spectrum. We are first and foremost: pro-democracy.

"Why We Fight

"Our mission is to protect the American Republic from Donald Trump and those who identify (publicly or privately) as MAGA supporters. While we are optimistic about the future, we are not complacent. In 2020, it took an as hoc coalition to ensure Trump's defeat. In 2022, we built a coalition across the political spectrum and again brought democracy some breathing room. We agree with Trump on only one thing: 2024 is a battle for America's future.

"President Abraham Lincoln led the United States through the bloodiest, most divisive and most decisive period of our nation's history. He fought, not because he wanted to, but because he knew the dual goals of preserving the Union and the end of slavery would be achieved only through armed conflict. At Gettysburg, he implored us not to forget those who had 'their last full measure of devotion' to preserving the American experiment.

"Today, we find our nation divided again -faced with a growing authoritarian movement populated by millions of radicalized voters, for whom this fight is existential as well as hyper -partisanship and Ultra -MAGA movements within the government itself. Donald Trump and those who ascribe to Trumpism are a clear and present danger to the Constitution and our Republic.

"This is why we fight."

This website is clearly passionate, devoted to the "defence of democracy". They also see "Trumpism" as an "authoritarian movement", one which is "a clear and present danger to the Constitution and our Republic". In fact, that which they refer to as "Trumpism" is not authoritarian, but far more dangerous. It is fascism.

As they are "dedicated to the preservation, protection and defence of democracy", those of us who are on the "Left", can unite with them, on that particular goal. After all, they have "expanded to include individuals from across the political spectrum". That includes those who consider themselves to be socialists, democrats, social democrats and even Scientific Socialists, which is to say Marxists, Communists. Yet in order for us to work with them, we must be able to put forward our own beliefs. If that is not possible, then there is no basis for working together.

The "common people", as the members of the working class refer to themselves, have to be made aware that this is indeed *fascism,* and not "Trumpism", or "MAGA", or even "authoritarianism". It is the same Nazi belief that our parents and grandparents fought against, in WW2! Trump is not the problem! MAGA is not the problem! The problem is the ruling class of billionaires, the bourgeoisie, those who "cannot rule in the old way"! They have decided to "change their method of rule"! The "new method of rule" is that of fascism! Trump is merely their figurehead!

As documented in a previous article, the billionaires have hired hundreds of loyal servants, in order to "screen" thousands of people, with the goal of placing those people in key positions, within the- soon to be created- Trump administration. Fascists one and all!

The threat is both grave and urgent. Feel free to take Trump "at his word", when he stated his intention to scrap the Constitution, "at or before" the next federal election! He means it! Join both mainstream political parties! Both serve the same class! Do not make a gift of the Republican Party, to the billionaires! Work with all the organizations which oppose fascism! Regardless of how Right Wing they may be!

For those of us who are Scientific Socialists, we can add another slogan, to our banners and posters:

Death To All Fascists!

CHAPTER 9

USE THE TWELFTH AMENDMENT AGAINST THE FASCISTS

All journalists, including those who work for the mainstream news outlets, are now deeply concerned with the fascist rhetoric of Trump. He recently announced his plan, to be put into effect "as soon as he is re-elected", to use the "Insurrection Act of 1792", as a legal pretext to send American troops into "blue cities". These include New York City and Chicago, which he referred to as "crime dens".

For the benefit of those who are not familiar with American politics, the political pundits have separated all the cities and states, within the country, into two categories. Some of them are classified as "Blue", or "Liberal", or "Democrat Leaning", as opposed to "Red", or "Conservative", or "Republican Leaning".

Allow me to stress the fact that these classifications are not mine. I mention this, only as a means of understanding, that which the journalists are saying.

This is the speech that Trump gave, which has caused so much concern: "The next time, I am not waiting. One of the things I did was let them run it and we are going to show how bad a job they do. Well we did that. We do not have to wait any longer."

This stilted jargon of Trump may be difficult to understand, as his command of the language is equivalent to that of someone with an elementary school education. It was his way of saying that while he was serving as President, he "let them" -meaning the democratically elected officials of various cities and states- "run it", meaning he let those officials perform their duty-, and in his opinion, they did a " bad job". For that reason, "next time", meaning as soon as he is once again President, he will not "wait any longer". Trump plans to take over those cities and states, once he is re-elected!

As Trump does not write his own speeches, merely reads from a teleprompter, we can only assume that his writers "dumb down" those speeches, writing in a manner that a twelve year old can recite. Now if only Trump could learn to read at that level!

Yet as one journalist, a Leftist fellow, pointed out in his broadcast:

"Experts are now starting to sound the alarms about Donald Trump's plans to invoke the Insurrection Act, should he re-take the White House in 2024…in order to 'crack down on those crime dens', as he calls them, which are liberal controlled cities.

"So now he is not talking about invoking The Insurrection Act to shut down on protests or riots, or even to crack down on crime. He is literally talking about using the Act to take over liberal cities, because they are so 'filled with crime'….. Not reviewable by courts, it does not take approval from Congress, cannot put an end to it …He can send military anywhere and take over, no guard rails, effectively Martial Law"

It is clear that this journalist is deeply concerned that Trump, once re-elected, plans to use the Insurrection Act to "take over". As for his statement that there are no "guard rails", I am deeply skeptical. No doubt both the courts and the Congress has a different opinion on that subject!

Even a former federal judge has given an opinion, concerning the "growing threat of Trump", as he phrased it. In his opinion, "conservative lawyers" should band together, in opposition. As he stated, "Should Mr. Trump return to the White House, he will arrive with a coterie of lawyers and advisers who, like him, are determined not to be thwarted again. The Federalist Society, long the standard bearer for the conservative legal movement, has failed to respond in this period of crisis. More alarming is the growing crowd of grifters, frauds and con men willing to subvert the Constitution and long established Constitutional principles for the whims of political expediency. The actions of these conservative Republican lawyers are increasingly becoming the new normal. For a group of lawyers sworn to uphold the Constitution, this is an indictment of the nation's legal profession. Any legal movement that could foment such a Constitutional abdication and attract a sufficient number of lawyers willing to advocate its unlawful causes, is ripe for a major reckoning. We must rebuild a conservative legal movement that supports and defends the American democracy, the Constitution and the rule of law and that incentivizes and promotes those lawyers who do the same."

Well spoken! As they are so concerned with their democratic rights and upholding the Constitution, perhaps they should do just that! In particular, they should pay strict attention to the Twelfth Amendment to the Constitution. As it is so important, I have decided to copy the whole thing:

"Twelfth Amendment

"The Electors shall meet in their respective states and vote by ballot for President and Vice-President, one of whom, at least, shall not be an inhabitant of the same state with themselves; they shall name in their ballots the person voted for as President, and in distinct ballots the person voted for as Vice-President, and they shall make distinct lists of all persons voted for as President, and of all persons voted for as Vice-President, and of the number of votes for each, which lists they shall sign and certify, and transmit sealed to the seat of the government of the United States, directed to the President of the Senate;–the President of the Senate shall, in the presence of the Senate and House of Representatives, open all the certificates and the votes shall then be counted;–The person having the greatest number of votes for President, shall be the President, if such number be a majority of the whole number of Electors appointed; and if no person have such majority, then from the persons having the highest numbers not exceeding three on the list of those voted for as President, the House of Representatives shall choose immediately, by ballot, the President. But in choosing the President, the votes shall be taken by states, the representation from each state having one vote; a quorum for this purpose shall consist of a member or members from two-thirds of the states, and a majority of all the states shall be necessary to a choice. [And if the House of Representatives shall not choose a President whenever the right of choice shall devolve upon them, before the fourth day of March next following, then the Vice-President shall act as President, as in case of the death or other constitutional disability of the President.–]The person having the greatest number of votes as Vice-President, shall be the Vice-President, if such number be a majority of the whole number of Electors appointed, and if no person have a majority, then from the two highest numbers on the list, the Senate shall choose the Vice-President; a quorum for the purpose shall consist of two-thirds of the whole number of Senators, and a majority of the whole number shall be necessary to a choice. But no person constitutionally ineligible to the office of President shall be eligible to that of Vice-President of the United States."

May I suggest that all "conservative lawyers" take note. Within the Twelfth Amendment, there is no mention of any political parties. That includes Republican and Democratic. Nary a word. Nor is there any mention of any "running mate", or any District, or any "popular vote". It does clearly state that the "Electors shall ….vote by ballot for President and Vice President…in distinct ballots". May I further suggest that it is the *Electors who vote for President and Vice President, of their choice! The states have nothing to say about this!*

In the interest of "practicing what you preach", feel free to make those arguments before the Supreme Court. In this way, you can "defend American democracy, the Constitution and the rule of law". Demand that the next American federal election *follow the Twelfth Amendment to the Constitution!*

In this way, the states will *not* be allowed to force any Elector to vote for any particular candidate, for President or Vice President! The Electors will be allowed, by Constitutional law, to vote for the candidates of *their choice!* Such candidates may, or may not, be the candidates of one of the two mainstream political parties. That is entirely at the discretion of the Electors!

As a bonus, the country will be spared the farce of a "popular vote"! Completely meaningless!

On a serious note, I can even suggest that "Leftist" lawyers may even assist, in this noble endeavour. Very likely, they will be only too anxious to help. Certainly a worthy cause!

Bear in mind that we do not have to like each other, in order to work together. We just have to respect each other, or at least tolerate each other, temporarily.

As for those who cannot imagine "conservative Republican lawyers" working with "socialist", or still worse, "Communist" lawyers, may I suggest that "stranger things have happened"!

On February 17, 1918, while Soviet Russia was still at war with Germany, Lenin met with a French officer, a demolition expert, also a self described "royalist, a champion of the French monarchy". Even though they hated each other, they had a common enemy, and were able to put aside their differences. After all, it was only temporary.

To borrow a sports metaphor, "keep your eye on the ball"! This is to say that the fascists must be stopped! The Republican Party is currently being used, by the billionaires, in order to promote fascism! Trump is merely their figurehead!

Focus on opposing the fascist movement, of the billionaires, by neutralizing the Republican Party! This can best be accomplished by forcing all federal elections, to abide by the Twelfth Amendment! The Republican Party can put forward any candidate of their choice! Trump or anyone else! It matters not! It is meaningless!

Under the Twelfth Amendment, no Elector will be forced to vote for Trump!

CHAPTER 10

CONCERNING THE "RANK AND FILE" MEMBERS OF MAGA

Numerous journalists, of the mainstream press, are deeply concerned that Trump remains the front running presidential candidate for the Republican Party. No one else is even close!

Yet there are certain "progressive" news outlets, on the internet, which are actively opposing Trump. It is to their credit, that their journalists have spoken to a great many "MAGA" supporters, among the "rank and file", the common people, those who support Trump. The results of these interviews have shocked those same journalists. In fact, the responses are most instructive.

From the responses to questions from the journalists, it is quite clear that the most devoted, "rank and file" supporters of Trump, appear to be honest, law abiding, hard working, tax paying, patriotic Americans. It is equally clear that they are among the less advanced members of the working class, the proletariat.

For that reason, it is necessary to *raise* their level of awareness. They are *not* the enemy! They have been *misled*, by the monopoly capitalists, the billionaires! They believe the lies they have been told, by certain members of the press! Many of them think that it is "against the law" to lie, on a news broadcast, or in a newspaper! So whatever the journalists say, must be true!

It is certainly to the credit of the progressive news outlets, that they have documented this. Yet the fact that those same outlets tend to *mock* those common people, is completely *unacceptable*! The beliefs of all common people should be *respected*! Such mockery must be *condemned*! It will serve only to drive ever more common people, into the ranks of MAGA!

Instead of mockery, may I suggest a different approach, one which is perhaps unconventional, as it has never been tried before, but one which promises a great deal. Allow me to explain.

Many years ago, a "celebrity", an actor whose name I cannot recall, mentioned something which I have never forgotten. Although I cannot recall the exact words, he said something to the effect that, "Once someone becomes a 'celebrity', two very strange things happen. The first strange thing is that people ask you questions, concerning a subject upon which you know nothing. The second strange thing is that you begin to answer those questions!"

The point is that common people listen to "celebrities"! For that reason, my suggestion is that progressive "celebrities" make videos, and place them on the internet. On these videos, feel free to explain a few things. Among other things, let your fans know that there is no law against lying. Everyone, including journalists, can lie, on the air, in public, and even publish it, in a newspaper. This happens, on a regular basis. As long as it is not slander, they can get away with this. You can then explain the difference between lies and slander.

It is also a fact that classes exist. Working class people are not aware of this, not class conscious. It has to be explained to them, as conditions of life, within the working class, do not lead to that awareness.

Let them know that the people who own almost everything of any considerable value, including the factories, mills, mines, railroads, airlines, shipping companies, banks, news outlets and even the internet, are members of a class of people, whom we refer to as monopoly capitalists, billionaires. They are technically referred to as the "bourgeoisie". They are also referred to as "imperialists". That is the class of people who are currently ruling the country.

Then there are working class people, common people, those who are forced to work by the hour, referred to as proletarians. The capitalists make their profit from our labour power. The less they pay us, and the harder they work us, the higher their profit. The more they pay us, the less their profit. It is clear that our interests are diametrically opposed. That which is in the best interests of one class, is in the worst interests of the other class. This is politely referred to as "class conflict". It is more accurately referred to as "class warfare".

It was Marx and Engels who first conducted a *scientific* examination of capitalism, and explained this quite clearly, in the Communist Manifesto. It is highly recommended that working class people read this article.

It was at the beginning of the twentieth century, that capitalism reached its highest stage, that of monopoly, or imperialism. At that time, it was Lenin who conducted an investigation into imperialism. He published his conclusions in his excellent work, Imperialism, the Highest Stage of Capitalism. Well worth reading!

As is well know, the common people of Russia, both workers and peasants, revolted in February of 1917, and overthrew their Emperor, or Czar. At that point, the Russian capitalists seized power.

At that time, Lenin was in exile, in Switzerland, but was able to return to Russia, in April of that year. He then wrote State and Revolution, in preparation for the approaching Russian Scientific Socialist Revolution, against the capitalists. This took place on October 25, old style calendar, or November 7, new style calendar, 1917. The first Scientific Socialist Republic was established, in the form of the Dictatorship of the Proletariat.

Now another revolution is approaching, an American Revolution. Here too, it is necessary to overthrow the American imperialists, the billionaires, and crush them, under the Dictatorship of the Proletariat.

The billionaires plan to set up a fascist, or "Nazi", dictatorship, with Trump as the "boss", scrap the Constitution, build huge concentration camps, imprison all who oppose them, deport millions of "undocumented" people, even American citizens, and deal "harshly" with the "vermin" who oppose them.

The only way to stop them is through revolution. After the revolution, they must be crushed, under the Dictatorship of the Proletariat.

The preceding was a mere suggestion of "talking points" which can be given, on the internet, by a "celebrity", or better yet, a group of "celebrities"! The more famous, the better! Common people will listen to such people! As long as these actors are properly prepared, and the presentation is well delivered, it will be most persuasive! As well, the career of the actors -actresses?- is sure to sky rocket! The same is true for those who write the scripts!

It may be necessary to conduct a series of videos, with each one touching on a different subject. The more popular the celebrities, the better! As long as the performances are entertaining, the working people will love it! Education through entertainment!

I can only suggest that all performances be based upon current events. Working people watch the news, quite closely. They pay strict attention! This gives performers a chance to point out the hypocrisy!

As an example, currently in the news, is talk of Section 3 of the Fourteenth Amendment.

Some journalists maintain that this Section prohibits Trump from once again, running for the presidency. Other journalists maintain that such is not the case. This can lead only to mass confusion, as both cannot be right!

Allow me to start by giving a brief summary of the Section in question:

"Section 3, No person shall be…president…(who)…shall have engaged in insurrection or rebellion, or given aid or comfort to the enemies thereof."

From that, most common people may get the impression that anyone who has engaged in "insurrection or rebellion", against the American government, is not allowed to run for president. I refer to this as good, old fashioned, "common sense".

It may perhaps come as no great surprise, to many readers, to find that there are a great many people, bourgeois intellectuals, one and all, who are of a different opinion. They think that it means nothing of the sort! They are completely devoid of common sense!

These "legal eagles" are of a different opinion, so that I offer here one argument, by one lawyer, put forward, on behalf of Trump:

"Section Three does not apply, because the presidency is not an office 'under the United States,' the president is not an 'officer of the United States,' and President Trump did not take an oath 'to support the Constitution of the United State".

At the time Trump was being sworn in as President, he took an oath to "preserve, protect and defend the Constitution of the United States". Yet the lawyers for Trump would have us believe that there is a difference between "support" and "preserve, protect and defend"! These lawyers get paid, quite handsomely, to argue these legal "technicalities"! All in the interest of placing Trump back in the White House!

All actors, writers and entertainers are artists, and I am certainly not about to advise any artist, as to how they should perform. Let the experts do that which the experts do best! Do not tell a mechanic how to "pull a wrench"! Do not tell a carpenter how to "swing a hammer"!

On the other hand, I can state certain facts. Aside from that which was previously mentioned, the billionaires are determined to wipe out the middle class. All of their "assets" are to be "appropriated" by those same billionaires. Their lust for wealth and power is unlimited! These people must be stopped!

That calls for revolution. The only successful revolution, is one that follows the advice of Marx! This is referred to as Scientific Socialism. That is one which calls for the Dictatorship of the Proletariat. This is only possible, if the proletariat is *class conscious!* It is up to progressive, middle class intellectuals, to bring them that awareness.

Middle class intellectuals have nothing to lose, and everything to gain, by becoming involved in the revolution. Those who assist in raising the level of awareness, of the proletariat, will perform a most valuable service.

After the successful Scientific Socialist Revolution, under the Dictatorship of the Proletariat, the contribution of these artists, writers and actors, will not be forgotten. On the contrary, they can expect to be richly rewarded. As an added incentive, their services will be in big demand.

The alternative is financial ruin, and a fascist dictatorship, with Trump as the dictator. Choose wisely.

CHAPTER 11

CONCERNING ATTEMPTS TO ESTABLISH A THIRD POLITICAL PARTY

I t is now clear that the revolutionary motion, currently sweeping the country, has even reached into the "highest reaches" of the middle class, the petty bourgeois.

That which has been referred to as the "Cult of Trump", is causing ever more concern, among the "upper middle class". Even those who consider themselves to be very "Right Wing", on the political spectrum, are deeply worried about this "political polarization", as they refer to it.

No doubt, all readers are familiar with the term "cult", although many may not be aware of the exact meaning. According to the internet, it is a rather derogatory term for a group of people, who are "led by a charismatic and self-appointed leader, who excessively controls its members, requiring unwavering devotion to a set of beliefs and practices".

The fact is that Trump is certainly "charismatic". At least superficially, he appears to be quite charming. But beneath the surface, is a different story. It is also a fact that he exercises great control over his followers, those who are referred to as, "MAGA" people, in that MAGA stands for Make America Great Again. Further, Trump demands -and expects!- unquestioning loyalty and obedience!

The middle class intellectuals have a point, when they refer to the "MAGA" people, as being part of a "cult". And not just any cult! A supremely dangerous cult! A threat to American democracy!

It is clear that Trump is also a demagogue. For those who are not familiar with the term, a demagogue is defined as "a political leader who seeks support by appealing to the desires and prejudices of ordinary people rather than by using rational argument."

Not only is Trump a cult leader and a demagogue, he is a highly successful one!

Truly, as Lenin stated, "Demagogues are the worst enemies of the working class"!

One bourgeois journalist maintains that his "followers do not care that he has been accused of ninety one felonies". He went on to say that most of his followers believe that "Trump tells the truth, rather than their minister, or even their family. He has a loyal following".

In an opinion essay, published in one of the most highly respected newspapers in the country, a conservative intellectual sounded truly pessimistic, writing in despair, that a "Trump dictatorship is increasingly inevitable! The more criminal charges he faces, the more popular he becomes!"

To this, I can only respond that such a dictatorship is certainly *not* "increasingly inevitable"! Such a threat calls for a proper *response*, on the part of progressive people, in order to forestall that threat!

I have documented in a previous article, the best way to combat the "Cult of Trump", is to raise the level of awareness of his followers, most of whom are common people. There is no need to repeat it here.

Another journalist, speaking from the depths of bitterness and frustration, bemoaned the fact that, as he stated, "People no longer respect the rule of law, or the Constitution. They have walked away from those thing, they do not trust any of the government entities. We are in a really scary place as far as our democracy is concerned. When people stop trusting in the rule of law, when people don't believe anybody but Donald Trump! What he tells them is the truth no matter what it is, even when they can see with their own eyes, that it is not true! Believe what I tell you, Trump controls the GOP base. Everyone is concerned with not antagonizing Trump".

My dear bitter journalist, if by "people", you are referring to the working people, common people, as "no longer respecting the rule of law, or the Constitution", you are so right! There is a reason for this! Common people are not entirely stupid! They are well aware that the "rule of law" and the "Constitution", apply *only* to working people! *Not* to the billionaires! The billionaires can break the law and violate the Constitution, as they see fit, and *get away with it! The law does not apply to the billionaires!*

Trump is currently facing *91 felony charges!* Yet he is a *free man!* If any member of the working class was facing even a *fraction* of those charges, that individual would be locked up! Never again to see the light of day! And *all* common people are well aware of this! So what did you expect?

As well, a former GOP Member of Congress wrote a book, in which she expressed herself in more entertaining terms: "He's told us what he will do. It's very easy to see the steps he will take. People who say, well, if he's elected, it's not that dangerous because we have all these checks and balances. They don't fully understand the extent into which the Republicans in Congress today have been co-opted. One of the things we see happening today is the sort of sleepwalking into dictatorship in the United States."

I must say, I am impressed! Someone certainly has a "way with words"! To refer to the MAGA movement as "sleepwalking into dictatorship" is one way of putting it! Either she is a talented writer, or possibly she hired a professional. I believe they are referred to as "ghost writers".

There are other middle class intellectuals, who have also expressed warnings. As I consider them to be instructive, I have chosen to mention a few.

A rather ominous warning comes from a journalist, writing in one of the most respected, "conservative" newspapers, in the country: "As perilous as it is for Republicans to say a negative word about Trump today, it will be impossible once he has sewn up the nomination. The Party will be in full general election mode, subordinating all to the presidential campaign. There will be no more infighting, only outfighting; in short, a tsunami of Trump support from all directions."

A former GOP governor spoke from a sense of deep bitterness, but accurately: "There is no Republican Party, there is a cult around Donald Trump. They did not adopt a platform in 2020, which means they did not set out what they stood for. It's what ever Trump tells us what we stand for today, that is what we shall stand for. So it is not a Party as we have known it. And believe what he says, he says what he is going to do. Believe him. He will tear down the very guard rails of our democracy. The rule of law, the respect for the Constitution, he means what he says. No one should question that."

Another bourgeois journalist supplied his own analysis: "There are still people who believe in his ability to 'drain the swamp'. This is 'slow walking' us to dictatorship. On the democratic side, we have Biden, who is old school. Young voters do not buy into this. The bigger problem is that there is an overall challenge to our capitalist society. A lot of people are not getting that American dream."

Finally! A bourgeois journalist who blurted out the truth! The truth is that "there is an overall challenge to our capitalist society"! Especially among the "young voters"! But what is the "American dream"?

According to the internet, it is the "ideal that the United States is a land of opportunity that allows the possibility of upward mobility, freedom, and equality for people of all classes who work hard and have the will to succeed."

There may have been a basis for such optimism, many years ago, but not now! Those days are long gone!

The key word here is *"classes"!* The members of the *working class,* have no "possibility of upward mobility, freedom and equality", if for no other reason than that there are *no* jobs, upon which to "work hard"! Make no mistake, the "young voters" are supremely well aware of this!

The preceding was a mere sample of middle class intellectuals, who have *identified* the threat posed by Trump, yet have chosen to do *nothing,* other than whine about that threat!

By contrast, in response to this "Cult of Trump", this "threat to democracy", there are numerous middle class intellectuals, who have decided to take action. They tend to refer to themselves as proud "Ultra Right Wing" conservatives. Many of them, but by no means all, are members of the Republican Party. They took part in setting up various conservative organizations, in order to oppose the "Cult of Trump". They think that a new political party will solve the problem. The results of their efforts are most instructive.

The first to take shape was the "Serve American Movement", (SAM), a political organization founded in 2017. Their stated goal is to "pass the elections vote limit or use petition gain access in other states, in order to contest future elections".

The precise meaning of their stated goal, is not clear. Yet the fact that they are making some sort of effort, is significant. At least, it is a step in the right direction.

Then in 2021, the "Renew America Movement", (RAM), was founded, reportedly by "former staffers from the Republican Party administrations", with the intention of working to "reduce political polarization" in the United States. Their stated goals were to unite Democrats, Republicans, and independent voters "... who have the courage to stand up against political extremists across the country".

Outstanding! Now that is certainly a worthy goal! To unite people, voters "who have the courage to stand up against political extremism", deserves our complete support!

As for "reducing political polarization", that is the very goal of every Scientific Socialist! This "political polarization" is merely the result of class struggle! As long as classes exist, we will have class struggle, or "political polarization", as she phrases it! It is the goal of all Scientific Socialists, true Marxists, Communists, to "reduce" this class conflict! This reduction can only happen *after* the Revolution, *after* the billionaires are overthrown and crushed, under the Dictatorship of the Proletariat! At that time, under Scientific Socialism, classes will continue to exist, so that the class conflicts will continue to exist! Yet they will be "reduced", as classes are gradually eliminated! Then, in the absence of classes, class struggle, or "political polarization", will also cease to exist!

That was not the only organization to take shape, in response to the MAGA movement, within America.

In 2021, The Forward Party, also known as Forward, (FWD), was created. It is a self described "centrist political party in the United States". The party, describes its goals as the "reduction of partisan polarization", and implementing "electoral reforms". As well, it has the goal of providing an alternative to the two major American political parties. It also states that, for the time being, candidates affiliated with the organization will remain members of the two major American political parties and America's third parties, as well as independent candidates.

As for the readers who may find that to be a bit confusing, bear in mind that it was copied from their website, on the internet.

Then in 2022, both the SAM and RAM merged with the Forward Party, using the Forward Party name. Now the stated goal of the Party is also to "break through a dysfunctional two party system that catered to the ideological fringes".

To describe the "two party system" as "dysfunctional", is completely accurate. It is absolutely dysfunctional! But to say that those two Parties "catered to the ideological fringes", makes no sense whatsoever! Unless the "ideological fringes" is a reference to the billionaires! That is the class of people to whom they "cater"!

Under the new alliance, Forward announced that there would be no traditional platform. Within each state, individual Forward Parties would determine their own priorities. Individual candidates, for that Party, would then develop their own priorities, as well as their own policies, based on those priorities.

One middle class journalist is of the opinion that The Forward Party is making a "dangerous miscalculation. It is betting that what a party opposes is more important that what it stands for". The idea is that "political beliefs no longer matter in a Party", so long as everyone is committed to moving, "Not Left. Not Right. Forward."

As such a policy can lead only to mass confusion, that journalist is no doubt correct! It is a step back from the stated goals of RAM!

In particular, the former Governor of New Jersey, and Forward Party Co-Chair, has been very vocal. Strangely enough, even though she is the "Co-Chair" of a new political party, she is still a member of the Republican Party.

In response to an inquiry, by a journalist, the former Governor of New Jersey stated,

"During the Trump era, I broke away from the GOP as it devolved into a cult of personality threatening democracy and our institutions; I served as the national spokesperson for the Republican organization, Renew America Movement, which worked to reduce political polarization.

"But Renew America was swept into the Forward Party in July 2022, one in a series of mergers and acquisitions of collapsing Republican reform groups that splintered off from the GOP in the wake of the Trump presidency. As they united under the new party's banner, the groups delivered their membership lists, resources and even their political professionals like me — despite having no ideological commonality."

This absence of "ideological commonality", among the "collapsing Republican reform groups", is a recipe for further collapse! It simply means that they have merged, without having any common goal! Her further experience confirmed this: "We had been promised our reform efforts would continue. Instead, I found an organization convinced it could maintain and grow its disparate coalition by not taking any positions at all. Its very existence was premised on the idea that, in the future, political parties will succeed by not having a philosophy of government, a shared vision or even a platform to unite behind. That's not what I signed up for."

I have deliberately gone into considerable detail, concerning the well meaning efforts, of middle class bourgeois intellectuals, to form an alternative political party. These are examples of how *not* to form such a Party!

This also shows how completely muddled these people have become! Bear in mind that some of them, or at least the former Governor of New Jersey, are among the most intelligent of the bourgeois intellectuals!

But as Lenin pointed out, in reference to a statement made by the British Prime Minister, one which made no sense: "This argument shows in particular how muddled even the most intelligent members of the bourgeoisie have become and how they cannot help committing irreparable errors. That, in fact, is what will bring about the downfall of the bourgeoisie."

Even the "most intelligent members of the bourgeoisie" are once again "muddled", once again "committing irreparable errors", and this is sure to bring about their "downfall"!

May I suggest, to progressive middle class intellectuals, that we should first face the fact that we live in a *class* society. Capitalists and workers, bourgeois and proletarians. Any political Party we create, can serve *one* of those two classes, or the other, it *cannot serve both!*

The two mainstream political Parties we now have, Democrat and Republican, serve the *same class!* The *capitalists!* The *billionaires!* The *bourgeoisie!* There is *no one Party serves all!*

As that is the case, there is only one Party that can serve the working class, the proletariat. That is the Party which embraces Scientific Socialism, the Party of true Marxism, the Party which calls for the Dictatorship of the Proletariat, the *touchstone* of a true Marxist, according to Lenin.

Such a true Communist Party can be *created* only by middle class intellectuals, although the most advanced workers can no doubt be of assistance.

The Communist Manifesto provides for this: "entire sections of the ruling class are, by the advance of industry, precipitated into the proletariat, or at least threatened in their conditions of existence. These also supply the proletariat with fresh elements of enlightenment and progress."

It is safe to say that the plan of the billionaires, to set up Trump as the head of a fascist dictatorship, qualifies as "threatening the conditions of existence", of the middle class. As well, all banks but eight, and all businesses but five, are "Too Small To Succeed"! They are destined for bankruptcy! As these businesses dissolve, the middle class will also cease to exist.

In a previous article, I have given practical suggestions, concerning the procedure to be followed, in setting up a true Communist Party. Suffice it to say that it is primarily up to middle class intellectuals, those familiar with the revolutionary theories of Marx and Lenin, to create such a Party. Be discrete, use the internet.

I can only add, that the need for a true Communist Party, one that calls for the Dictatorship of the Proletariat, is becoming ever more urgent.

Such a Party must, of necessity, be limited to a rather small number of dedicated, disciplined, professional revolutionaries. It must be highly centralized, with a well established Party platform, so that everyone knows the policy of the Party. The key lies in "democratic centralism", an organizational principle, in which decisions are taken after discussions, among all Party Members. Once the general Party lines are established, no further discussion is permitted. Factions are not allowed. No member or institution is allowed to disagree on a policy, after it has been agreed upon by the governing body of the Party. The penalty for doing so is expulsion from the Party.

Bear in mind, that the billionaires plan to set up a fascist dictatorship, with Trump as the figurehead leader. The only way to stop them, is through a Scientific Socialist Revolution, which will result in the Dictatorship of the Proletariat.

For that reason, the level of awareness of the working class must be raised. The proletariat, or at least the most advanced workers within the proletariat, must be raised to the level of Scientific Socialists. This can only be accomplished, through the efforts of a true Communist Party.

The formation of that true Communist Party, Dictatorship of the Proletariat, is now in the hands of class conscious, middle class intellectuals. I want you to know, that I have complete confidence in you.

CHAPTER 12

COUNTERING PREPARATIONS FOR A FASCIST DICTATORSHIP

The latest details, concerning preparations for the fascist dictatorship, have recently come to light. The journalists have managed to get their hands on a questionnaire, reported to be "created by Trump allies", to be used to "determine the suitability of people who can serve in the next Trump administration."

The purpose of the questionnaire, is to select "suitable" people, for the "bureaucracy", within the anticipated Trump administration. These people must be loyal *only* to Trump, devoted followers, prepared to cater to his every whim. Such people are known as "sycophants", commonly referred to as "yes people".

In the interest of bringing all members of the working class "up to speed", I should mention that the technical definition of a sycophant, is "a person who acts submissively towards someone important, in order to gain advantage".

The question that has horrified so many people, is the following: "The president should be able to advance his own agenda, through the bureaucracy, without hindrance from unelected federalist officials. Agree or disagree."

This mention of "unelected federalist officials", is a none too subtle reference to the American judicial system! All federal judges, including those on the Supreme Court, are *appointed*! They are *not elected*! They are part of the judicial system, which serves to "act as a brake" upon the "power of the executive branch" of government! Meaning the President! This is commonly referred to as a "guardrail" of democracy! Yet Trump wants to "advance his own agenda, without hindrance", from those judges!

The only way this can happen, is by scrapping the Constitution! This is the very thing Trump plans to do, on his *first day* in office! He even *stated* that he plans to "*act as a dictator*", on his "*first day in office*"!

His former personal lawyer, who says he "knows Trump better than anyone else", says that Trump is not joking! In his opinion, on his *second* day in office, Trump will declare himself to be king, or "Fuhrer", and demand everyone swear allegiance to him, and even give a straight arm Nazi salute!

A great many middle class Americans, those who consider themselves to be "proud conservatives", are now actively opposing this attempt to establish a fascist dictatorship. This is an indication of the strength of the revolutionary movement. It has reached even into the highest levels of the middle class.

For that reason, a very conservative lawyer for the American Bar Association has convened a "Blue Ribbon Task Force For American Democracy".

The goal of this Task Force is to "preserve and protect our democratic institutions". Further, it is reported that "numerous State Bar Associations, as well as the Urban League, and various professional people, are also involved."

Once again, in the interests of bringing the members of the working class "up to speed", I will mention that a "Blue Ribbon Task Force" is defined as "a group of exceptional people appointed to investigate, study or analyze a given question."

The fact that they refer to themselves as "exceptional people", is merely an indication of their high opinion of themselves! Such people tend to be supremely arrogant!

Even though these people are very conservative, or "Right Wing", they have taken a progressive stand. For that reason, those who are Scientific Socialists, Communists, can support them, on this particular issue.

It is reasonable to expect that working with these "exceptional" people, can be most tiresome. Perhaps those who are working with these middle class professional people, can take some comfort in the fact that they hate us, every bit as much, as we hate them!

The class struggle is not a "bed of roses"! There are times when it is necessary to "form temporary alliances", with those who are at best "unreliable", according to Lenin.

Then again, such support must be conditional upon being able to express our Marxist beliefs, including that of Revolution and the subsequent Dictatorship of the Proletariat. If that is not possible, then any alliance, however temporary, is out of the question.

This Task Force is of the opinion that "the judiciary is our last best hope for saving our democracy".

Such is hardly the case! Yet it is very likely that these "exceptional" middle class professional intellectuals, are not capable of understanding the class struggle! For that reason, they cannot possibly understand that our "last best hope" for "saving our democracy", is the working class, the proletariat!

It is necessary to *raise the level of awareness* of the proletariat, and any court challenge can help to serve that purpose. The press, especially the

"Leftist" press, which is very active on various web sites, must cover these court challenges, in great detail.

That being said, it is encouraging that the "exceptional members", of this "Blue Ribbon Task Force" are determined to fight the multi billionaires, as they try to establish a fascist dictatorship.

As that is the case, may I suggest that "the best way to kill a snake is to cut off its head"!

Half measures get us nowhere! Prepare to go straight to the Supreme Court, to defend the Constitution! Challenge the 2020 federal election, on the grounds that it did not follow the procedure laid out in the Twelfth Amendment! Further demand that the forth coming 2024 federal election, follow that procedure!

The ruling class of monopoly capitalists, the multi billionaires, have *subverted* the Constitution, through their creation of the "Two Party system". As a result of this -illegal!- Two Party system, the Electors may soon be forced to choose between two older men, both of whom may be suffering from "cognitive decline", commonly referred to as "dementia".

Mental illness is not a joking matter! The brain is an organ of the body, and it is subject to illness, as is any other organ! Just as no one should laugh at a heart attack, so too, no one should crack jokes about mental illness!

It is the Two Party system that is the problem! That is Unconstitutional! It is also *enabling* the multi billionaires to establish a fascist state! Or at the very least, they are making a supreme effort! They must be stopped!

By Constitutional law, it is up to the *Electors*, who are appointed by the states, to vote on one ballot for the person of *their* choice, for President, and on a separate ballot, to vote for the person of *their* choice, for Vice President. There is no mention, in the Twelfth Amendment, of any political Party, District, popular vote, or "running mate". Further, the states have *no* right to meddle in a federal election.

Under the Two Party system, a "popular vote" takes place, in which all American citizens are allowed to vote for the candidate for President, and his "running mate", of either the Democratic Party, or of the Republican Party. This popular vote is not mentioned in the Twelfth Amendment! It is a mere formality! It has no legal standing!

Then the votes of these citizens are counted, and based on the results of that popular vote, the Electors, of that particular state, are *forced* to vote for the President, and his "running mate", as Vice President.

This procedure is completely Unconstitutional, as the states have *no right to meddle in a federal election!*

If the multi billionaires have their way, at the time of the 2024 election, the Electors will be *forced* to choose between Biden and Trump. This is strictly Unconstitutional!

If the Two Party system is challenged, before the Supreme Court, no doubt it will rule that all federal elections must follow the procedure laid out in the Twelfth Amendment. At that point, the Electors will be allowed to vote for any candidate, of their choice, for the office of President, as well as Vice President. No state will be allowed to force any Elector to vote for any particular person. All state laws, which require an Elector to vote for the candidates of one of the major political parties, will be struck down as Unconstitutional.

Such a Supreme Court ruling is sure to "throw a monkey wrench" into the plans of the multi billionaires! Their plans, for a fascist dictatorship, relies upon the election of a candidate, of one of the two political parties, as President!

No doubt, among the "exceptional" people on this "Blue Ribbon Task Force", there are a great many lawyers. Those who are experts on Constitutional law, should prepare the arguments, to be presented before the Supreme Court. In that way, the next federal election can be legal, truly democratic, following the procedure laid out in the Twelfth Amendment.

It may be objected that not all members of the ruling class of monopoly capitalists, the multi billionaires, are determined to establish a fascist dictatorship, with Trump as the dictator. In fact, a "major conservative" organization, "Americans For Prosperity Action Group", has just endorsed the former Governor of South Carolina, Nikki Haley, as the Republican nominee for president. According to their web site, they state that they plan to "turn the page on the current political era". They go on to state that, "Trump and Biden had their chance, they can't fix what is broken, it's our time to choose a new leader who will unite our Party, our Nation".

The fact that this "Action Group" is endorsing Nikki Haley, is significant. The reason I say this, is because a multi billionaire is behind this Group! Perhaps the monopoly capitalists are "covering all the bases", if you will excuse the sports metaphor. Just in case the common people turn against both Biden and Trump!

Under a fascist regime, the figurehead dictator is not important. The important thing is to have a proper "state apparatus" set up. The fascists made this quite clear, in their statement: "It is not enough for conservatives to win elections. If we are going to rescue the country from the grip of the radical Left, we need both a governing agenda and the right people in place, ready to carry this agenda out on day one of the next conservative administration."

The "governing agenda" is to scrap the Constitution, on the first day of office, of the next President! They have already planned, in advance, the goals of the first six months of the anticipated fascist administration! This resembles more of a military campaign, than a political platform!

As that is the case, perhaps it would be best to respond to this attempt at establishing a fascist dictatorship, in much the same manner, as we would respond to a war. After all, it is a war!

The key to winning this war lies with the working class, the common people, the proletariat. We have got to raise their level of awareness, or at least that of the most advanced workers, to that of the level of Scientific Socialists, Marxists, Communists.

At the same time, we must not ignore the less advanced. We must prepare literature -and even videos!- for the common people, those who have embraced MAGA. Both literature and videos must be presented in a popular manner, without being condescending, and by no means vulgar. It is not the fault of countless common people, that they believe the lies of the monopoly capitalists, the multi billionaires!

In the case of videos, it would be best if Leftist celebrities, the more popular the better, presented the facts. In an entertaining manner, of course. That is where Leftist writers come in, with perhaps a little song and dance routine. That is entirely up to them. No doubt, the common people will love it. Education through entertainment!

In response to the multi billionaires, and their plan for a fascist dictatorship, we must have our own plan:

Prepare For the Dictatorship of the Proletariat!

CHAPTER 13

SCIENTIFIC SOCIALISM

A great many journalists, including those who work for the main stream media, are now expressing deep concern over the plans of the multi billionaires, to establish a fascist dictatorship in the United States.

Incidentally, those who work for the mainstream media, are referred to as "bourgeois journalists", as the billionaires own those news outlets. For that reason, those journalists have to "slant" the reporting, in their favour.

They frequently manage this, by first presenting the news, and then conducting an interview, with an "expert" on the subject. These "experts" then express their opinion, concerning the news which was just reported. All too often, this opinion is in direct contradiction to the facts. The idea is to confuse viewers. This is known as "spinning" the news.

In contrast to the bourgeois journalists, there are journalists who work for the independent media. These journalists were formerly referred to as "underground" journalists, as that was in the days before the internet. Now such people are referred to as "Leftist" journalists. Their reports tend to be more accurate, certainly not biased towards the billionaires. Yet their analysis too, can be faulty, as many of them are either not aware of the scientific theories of Marx and Lenin, or disagree with those theories.

These revolutionary theories are referred to as "Scientific Socialism". As so many working class people are just now becoming politically active, they deserve a little explanation. No doubt, middle class readers will find this tiresome, but my main concern is with the proletariat. After all, the success of the revolution, -as well as the defeat of the fascists!- depends upon the proletariat!

Our current system of capitalism, is rather recent. It came into existence with the industrial revolution, which took place between 1720 and 1760, in Great Britain, and quickly spread around the world. Before that time, everything was made by hand. Afterwards, hand crafted items became very rare and expensive, as they are now. A specialty product!

Historians are correct, when they state that the industrial revolution was the greatest thing to happen to humanity, since the domestication of plants and animals! It was absolutely revolutionary!

This also gave rise to two new classes. The "burghers" of the time, saw an opportunity to invest their money in factories, mills, mines and other "means of production", as well as banks and other "financial institutions", and make some "serious money", as they phrase it. At the same time, they invested in railroads and shipping lines, which they refer to as the "means of transportation", so that they can get the "raw materials" to the "point of production", as well as the "finished product", to market.

This is part of the stilted jargon the bourgeois economists use. The members of the working class, or at least the most advanced members to the working class, must learn to understand this jargon.

In the process, a remarkable transformation took place. The money, which the burgers invested, became known as "capital", and they became known as "capitalists". The name burgher became corrupted, into the name "bourgeois". The most wealthy of the bourgeois, became known as the "bourgeoisie". For that reason, the billionaires of today, are technically referred to as the bourgeoisie.

Yet no class can live in isolation. This is to say that a class of slave owners requires a corresponding class of slaves, while a class of nobility requires a corresponding class of "commoners", mainly peasants. The newly created class of bourgeois, or capitalists, needs a corresponding class of workers, or "proletarians", those who sell themselves, by the hour.

The bourgeois, or capitalists, needed workers, or proletarians, to run their machines. For that reason, they generally hired peasants, commonly referred to as farmers. In this manner, the members of the class of peasants, became transformed into a class of proletarians.

As the bourgeois became ever more wealthy, they decimated the class of peasants, the very class the nobility depended upon, for their wealth and power. The more wealthy the capitalists became, the more impoverished the nobility became.

This was not at all what the bourgeois had in mind! Their greatest ambition in life, was to join the nobility! Yet they accomplished precisely the opposite! They destroyed the nobility!

It is instructive to note that the earliest plans, of this newly created class of capitalists, did not work out.

I use this as an example of changes in methods of production, which lead to events, including the creation and destruction of different classes, which are beyond the control of "mere mortals".

Even the bourgeois economists could not explain how the capitalists were making their money. But then, the capitalists did not care! They only cared about their profit, or "bottom line", as they referred to it. That is true to this day.

No one could understand this, *except* Karl Marx! It was Marx who conducted a thorough, *scientific* examination of this newly created system of capitalism. He and Engels published those findings, in that which they referred to as the Communist Manifesto.

Further, in a letter, which Marx wrote, in 1852:

"And now as to myself, no credit is due to me for discovering the existence of classes in modern society or the struggle between them. Long before me bourgeois historians had described the historical development of this class struggle and bourgeois economists, the economic anatomy of classes. What I did that was new was to prove: (1) that the existence of classes is only bound up with the particular, historical phases in the development of production, (2) that the class struggle necessarily leads to the Dictatorship of the Proletariat, (3) that this dictatorship itself only constitutes the transition to the abolition of all classes and to a classless society."

It is important to bear in mind that this most important theory, that "the class struggle *necessarily* leads to the *Dictatorship of the Proletariat*", was proven by Marx, as early as 1852!

Lenin refers to this theory, the Dictatorship of the Proletariat, as the *"touchstone of a true Marxist"!* It is the worst nightmare of every billionaire! With good reason! No one wants to lower their standard of living! The billionaires are no exception! The very idea of losing all their wealth, of being "degraded", forced to "stoop" to manual labour, is too terrible for words! They literally cannot imagine this!

Yet the billionaires, as well as almost all members of the upper middle class, are well aware of this theory. The reason for this is quite simple. The parents, among the upper classes, make sure that their children attend University. It is only in University that the students are exposed to the revolutionary theories of Marx and Lenin.

This creates a little problem for certain of the parents, in that some of their children may not score top marks on the college entrance examinations. While money may not "buy happiness", it certainly buys almost everything else! That includes admission to the finest Universities in the country.

This became public knowledge recently, amid a recent scandal, which involved upper middle class parents. They were caught, paying someone to take the admission examinations, on behalf of their children. This made headlines, if only because the parents were *celebrities! Movie stars! Famous actors!*

The reason I mention this, is to drive home the point that *celebrities have power!* Such people may be professional athletes, but the vast majority are "Hollywood Movie Stars"! The more attractive, the more famous, the more powerful!

As for those who may object -quite reasonably!- that this makes absolutely no sense, I can only respond, that it does not have to make sense! This is just the way it is! Face it! Use it!

Working class people, common people, love certain actors! Or actresses, if you want to be persnippity about it! We can use this to our advantage!

We know, for a fact, that certain Hollywood celebrities are at least sympathetic towards socialism. This is commonly referred to as being "Leftist". The reason we know this, is because some of them have been arrested, while exercising their democratic right to peaceful protest.

May I suggest, to those celebrities, that there is more than one way to protest! Learn from previous experience! In the nineteen sixties, the country was also experiencing revolutionary motion. A number of Hollywood writers, no doubt "Leftists", came up with the idea of a "sitcom", a weekly television show, which *exposed* the class distinctions. This show was exceptionally popular. It was titled The Beverly Hillbillies.

That show was wildly successful! Working people loved it! The bourgeois critics *hated* that show, every bit as passionately as the working people *loved* it! The viewers could relate to the Hillbilly Clampetts, with their simple life style, as opposed to their neighbours, the multi millionaire bankers.

To all Leftist celebrities, whether Hollywood "Movie Stars" or otherwise, face the fact that you have power! It is dishonest to deny this fact. Spare all of us the false humility! Instead, *use* this power, in the service of the working class. The more fans you have, the greater the power. All of those fans will pay strict attention to whatever you say!

It is also a fact that we do not get to choose the people, with whom we work. We are stuck with them. Yet on our own time, we are free to associate with whomever we want!

For that reason, it is safe to assume that Leftist celebrities tend to "buddy up", with other Leftist people. These may include other celebrities, as well as writers. It is the writers who are important to the success of any presentation. Yet the "delivery" is critical!

The level of awareness of the working class, the proletariat, has to be raised! Certain expressions, such as Socialist Revolution, Soviet Power, Scientific Socialism and Dictatorship of the Proletariat, must become common place! Who better to take part in raising that level of awareness, than the Hollywood Movie Stars? The MAGA people listen to Trump, because he is a celebrity. Those same people will listen to other Hollywood celebrities, Movie Stars! Especially if those Movie Stars are most beautiful and charming! There is no shortage of Leftist people, in Hollywood, who fit that description!

For the benefit of Leftist people who think that socialism is merely a good idea, but not practical, may I suggest that you face the facts.

Over the years, a great many well meaning individuals, have made countless attempts to set up socialist societies, under capitalism. All have ended in failure. Countless others have come out with "alternative" theories of socialism. These are all referred to as "utopian socialism", as they are not based on the Revolutionary theories, of Marx and Lenin.

The only true Revolutionary theories, are the Scientific Socialist theories of Marx and Lenin. They are based on facts, not conjecture. The experience of the successful Russian Scientific Socialist Revolution, which gave birth to the Union of Soviet Socialist Republics, as well as the successful Chinese Revolution, proved the correctness of those theories.

Capitalism has been restored in both of those formerly socialist countries. The reason is that after every Scientific Socialist Revolution, classes continue to exist, and the billionaires make a supreme effort to restore their "paradise lost". That is the reason they have to be crushed, under the Dictatorship of the Proletariat.

In the case of the Soviet Union, as well as China, after the Scientific Socialist Revolution, the capitalists were not properly crushed, at least not with sufficient enthusiasm. A fundamental tenet of Marx, the Dictatorship of the Proletariat, was not properly carried out!

The leaders of those formerly socialist countries, Stalin and Mao, made mistakes, which I have covered elsewhere. There is no need to repeat it here. Suffice it to say that it is up to us to honour the memory, of our heroic ancestors, those great Revolutionaries, by learning from their mistakes.

Now to return to the subject of capitalism. *At first,* capitalism played a most revolutionary role. Of that, there can be no doubt! This is elaborated quite clearly, in the Communist Manifesto. But then, that was in the days of competitive capitalism.

That changed, quite dramatically, around the beginning of the Twentieth Century. At that point, capitalism reached the stage of monopoly. Instead of competition between capitalists, within each branch of industry, capitalists came together and agreed to set prices. As there is no competition, capitalism stagnated, as there are almost no great inventions. Monopoly stifles inventions, as well as competition.

Of course, there were other changes which took place, such as the export of capital, and not just goods. Once again, the capitalists knew that something had changed. Not that they cared! They just knew that they were making huge profits, and that was their only concern. They merely referred to this new state of monopoly capitalism as "imperialism".

It was Lenin, following in the footsteps of Marx, who conducted a proper *scientific* examination of monopoly capitalism, also known as imperialism, and published his findings in, Imperialism, the Highest Stage of Capitalism. In that work, he gave a brief definition of imperialism:

"Without forgetting the conditional and relative value of all definitions in general, which can never embrace all the concatenations of a phenomenon in its full development, we must give a definition of imperialism that will include the following five of its basic features:

1) "the concentration of production and capital has developed to such a high stage that it has created monopolies which play a decisive role in economic life;

2) "merging of bank capital with industrial capital, and the creation, on the basis of this 'finance capital,' of a financial oligarchy;

3) "the export of capital as distinguished from the export of commodities acquires exceptional importance;

4) "the formation of international monopolist capitalist associations which share the world among themselves and

5) "the territorial division of the whole world among the biggest capitalist powers is completed. Imperialism is capitalism at that stage of development at which the dominance of monopolies and finance capital is established; in which the export of capital has acquired pronounced importance; in which the division of the world among the international trusts has begun, in which the division of all territories of the globe among the biggest capitalist powers has been completed."

In that same article, he went on to say that "the political features of imperialism are *reaction* all along the line".

Our current crop of American imperialists, have found themselves "on the horns of a dilemma". They have reached a state of crisis. Their "time honoured" method of rule, the "Two Party System", is no longer working. For that reason, they are attempting to *change* their *method* of rule. The new method of rule, which they have decided upon, is that of fascism.

Under fascism, the precise dictator is of little consequence. He -or she!- is a mere figurehead. The important thing is that this individual serve the imperialists, the billionaires, the bourgeoisie.

The imperialists must be stopped! In the past, the fascists of two highly industrialized countries, Germany and Italy, were successful! That proves it can be done! Now is not the time to be complacent! Now is the time for action!

Even though a great many things should be done, it is necessary to focus on that which Lenin refers to as the "key link":

"Every question runs in a vicious circle because political life as a whole is an endless chain consisting of an infinite number of links. The whole art of politics lies in finding and taking as firm a grip as we can of the link that is least likely to be struck from our hands, the one that is most important at the given moment, the one that most of all guarantees its possessor the possession of the whole chain."

As is well known, it is my contention that the current "key link", is to raise the level of awareness of the working class, so that at least the most advanced strata of the proletariat, embraces Scientific Socialism, Soviet (Council) Power and the Dictatorship of the Proletariat.

The problem is that the working class is *not* aware of itself as a class, complete with its own class interests. That awareness must come to the working class, from an outside source. That outside source is middle class intellectuals.

The Communist Manifesto provides for this "outside source", which is destined to bring the class awareness to the working class:

"In times when the class struggle nears the decisive hour, the progress of dissolution going on within the ruling class, in fact within the whole range of old society, assumes such a violent, glaring character, that a small section of the ruling class cuts itself adrift, and joins the revolutionary class, the class that holds the future in its hands. Just as, therefore, at an earlier period, a section of the nobility went over to the bourgeoisie, so now a portion of the bourgeoisie goes over to the proletariat, and in particular, a portion of the bourgeois ideologists, who have raised themselves to the level of comprehending theoretically the historical movement as a whole."

Without doubt, the "class struggle" is indeed "near the decisive hour". The "process of dissolution" going on within the "whole range of old society", is now sharp and clear! Now we can expect a "small section of the ruling class", a "portion of the bourgeois ideologists", to "go over to the proletariat". They will bring with them the awareness of the revolutionary theories of Marx and Lenin.

It is entirely possible that "a small section of the ruling class" has already "cut itself adrift", and may even be preparing to "join the revolutionary class". That "small section" may include the members of the Americans For Prosperity Action Group, as they have endorsed Nikki Haley, as the presidential candidate of their choice, for the Republican Party.

Even though at least one billionaire is backing this Group, it is entirely possible that he, and a few others, have decided to oppose the plans of the ruling class of billionaires, to form a fascist state.

Perhaps they are serious, when they state that they plan to "turn the page on the current political era". They go on to state that, "Trump and Biden had their chance, they can't fix what is broken, it's our time to choose a new leader who will unite our Party, our Nation".

Either way, it is now necessary for "a portion of the bourgeois ideologists, who have raised themselves to the level of comprehending theoretically the historical movement as a whole", to "go over to the proletariat".

It is very likely that all Leftist Hollywood "Movie Star" celebrities, as well as writers, have a University education, and are aware of the revolutionary theories of Marx and Lenin. As the Universities are careful to teach only the bourgeois distortions of those theories, it is also likely that they are at best, deeply skeptical.

May I suggest that all middle class intellectuals, and not just the Hollywood celebrities, read the essential works of Marx and Lenin, with an open mind. Keep the thought uppermost in mind, that the ruling class of billionaires are in the process of setting up a fascist society, preferably with Trump as the figurehead dictator. It is up to the working class, the proletariat, to stop them.

What better way to oppose our class enemies, the billionaires, than with their own weapons? Just as the Socialist Soviet Union opposed the Nazi Blitzkrieg, at the time of the Great Patriotic War, with their own Soviet Blitzkrieg, so too I am suggesting that we oppose the Hollywood celebrity of the billionaires, in the form of Trump, with our own celebrity, or celebrities, preferably ones far more charming and beautiful.

Without doubt, the MAGA people will listen to these lady Movie Stars! At least, I hope they are ladies! Better yet, a mixed group of Movie Stars! The more the merrier! Especially if the guys are "Hunks"!

Although I write this in jest, the fact remains that it is true! The threat is real, and the solution, or at least part of the solution, lies with the Hollywood Movie Stars. All working people, including the less advanced, the MAGA people, will listen closely to that which the Movie Stars say. Strange but true.

The Hollywood Leftist writers, working behind the scenes, can prepare the script, which is sure to be entertaining, written in a popular manner. Then the Hollywood "Movie Stars" can make the presentation. The common people, including MAGA, will love this!

In conclusion, it is clear that the ruling class of billionaires is currently preparing to set up a fascist dictatorship, with Trump as the figurehead dictator. There is strong support for this, among the less advanced strata of the working class, the MAGA people. We have got to raise their level of awareness. One way to do this is through the use of celebrities, preferably Hollywood Movie Stars.

As soon as the vast majority of working people reject MAGA, and endorse Scientific Socialism, Soviet Power and the Dictatorship of the Proletariat, then the threat of fascism will be neutralized, and the success of the next American Revolution, will be assured.

CHAPTER 14

GROWING MIDDLE CLASS OPPOSITION TO FASCISM, LED BY TRUMP

The journalists who work for the mainstream press, are quite gleefully reporting that the Supreme Court has agreed to "expedite" the request of the "Special Prosecutor", to rule on the "Trump immunity claims". In fact, the Supreme Court gave the lawyers who represent Trump, until December 20, a mere week, to respond.

Those same journalists refer to this as a "big deal"! I could argue that this falls under the heading of "somewhat exceptional", rather than a "big deal", but then, that would be pointless. Besides, "simple minds are easily amused". Perish forbid that I should "rain on their parade"! Still, there is a reason for their excitement. This requires a little explanation.

If the common people are confused by the various charges, as well as court appearances, of Trump, it is completely understandable. It is confusing!

For the last two months, the press has been focused on his trial in the State Supreme Court of New York, in Manhattan.

In the case of that trial, the New York Attorney General, Letitia James, alleges that over a period of 20 years, "the Trumps grossly inflated the former president's net worth by billions of dollars", and of "cheating lenders and others with false and misleading financial statements".

This is a "civil lawsuit", so that no charges have been placed. This lawsuit accuses Trump and his firm, of "persistent and repeated business fraud". So the Attorney General is seeking a "$250 Million judgement from Trump, for defrauding lenders, and others". She is also asking the Court to "prohibit any of the Trumps leading a company in the state of New York".

Here we have a very clear cut example, of the "double standard" of the judicial system! The "authorities" are alleging that Trump and his family have been committing fraud for 20 years, and yet neither Trump nor any of his adult children, are facing any criminal charges! The Attorney General is seeking to have the judge "slap their hand", and perhaps even give them a good "tongue lashing"!

Instead of *charging* any of those billionaires with any criminal charges of fraud, *as is her duty,* she has resorted, in the finest tradition of the bourgeois, to "passing the buck"! According to press reports, she has "referred her findings to federal prosecutors in Manhattan", who could possibly "open a criminal investigation into bank fraud". Let someone else handle that political "hot potato"!

As I write this, the court has adjourned for possibly a month, with "closing arguments" scheduled for January 11, 2024. The judge is expected to "issue his ruling", several weeks later. That "ruling" will be concerned strictly with the amount of damages to be paid.

That is merely one civil court trial that Trump has to face. There are four other trials scheduled, which concern the felony charges against Trump. For that reason, these trials are referred to as "criminal trials", which can involve prison time, as opposed to civil trials, which do not.

As is well known, Trump is currently facing *91 felony charges!* By definition, a felony charge is quite serious, and can result in a rather lengthy prison sentence, if convicted. It is equally well known that Trump appears to be completely unconcerned, with the possibility of spending the rest of his life in prison. There is a reason for this. Trump is not the slightest bit worried about going to prison!

Trump is a billionaire! The laws do not apply to billionaires! Only to common people! The billionaires can do as they please, and get away with it. And they do! Trump is living proof of that!

Donald Trump is making American history! The journalists are practically in awe, as they report that, "over a five month span, former President Donald Trump was charged in four criminal cases. In Washington, D.C., he faces four felony counts for his (alleged) efforts to overturn the 2020 election. In Georgia, he faces 13 felony counts for his (alleged) election interference in that state. In New York, he faces 34 felony counts in connection with (alleged) hush money payments to a porn star. And in Florida, he faces 40 felony counts for (allegedly) hoarding classified documents after he left office and impeding the government's efforts to retrieve them".

American history has a long and distinguished list of famous individuals. These include Billy the Kid, Jesse James, Cole Younger, Wes Hardin, Machine Gun Kelly, Baby Face Nelson, Pretty Boy Floyd, Scarface Al Capone, Dutch Schultz, Sammy the Bull Gravano, and Teflon Don John Gotti. All of these people had one thing in common. They were leaders!

A great many people followed those leaders. To prison, to the gallows, to the grave! Yet none of those psychopaths could "hold a candle" to Trump! He is in a league of his own! He is not following in their footsteps! He is blazing his own trail! That is the trail to his own fascist dictatorship!

In an attempt to stop Trump and the fascists, the Special Prosecutor for the United States Department of Justice, Jack Smith, is "investigating Trump's (alleged) role in the January 6 United States Capitol attack and Trump's (alleged) mishandling of government records, including classified documents."

This has given rise to plans for three trials against Trump. The first is scheduled to take place on March 4, 2024, in the District of Columbia. In that trial, Trump is facing a "four count indictment", or four felony charges, to put it in simple English.

The second trial is scheduled to take place on May 20, 2024, in Florida. In that trial, he is facing "37 felony counts", or in simple terms, 37 charges of various felonies.

As well, in New York City, the borough of Manhattan has "indicted the former president on 34 felony counts related to (alleged) hush money payments to an adult film star". That trial is scheduled to begin sometime in March.

Lest we forget, Trump is also "facing possible charges from the top prosecutor in Fulton County, Georgia", concerning his (alleged) efforts to "reverse the outcome of the presidential election in the state".

All of that is in addition to "a separate civil case" against Trump, brought by the New York Attorney General.

No wonder the public is confused!

It is the plans for the criminal trial of March 4, that is creating such a commotion, among the mainstream journalists. The reason for this is that Trump has instructed his lawyers to stall all criminal trials, at least until after the next federal election, scheduled for November of 2024. At that point, he fully expects to become the President Elect, soon to be President.

Trump recently "doubled down" on his plan to declare himself "dictator on day one", of his presidency. Trump is not joking! The people who will fill all government posts, within his anticipated administration, are currently being selected. Their duty is to make sure that no one opposes the "Will of Trump". This is referred to as fascism. Fascist dictators do not worry about criminal charges!

Special Prosecutor Jack Smith is well aware of this. In an effort to combat this plan of the ruling class of billionaires, to set up a fascist society, with Trump as the figurehead dictator, Smith is taking action.

In particular, Smith has asked the Supreme Court to rule – urgently! – on whether Trump has legal immunity, as the ex-president claims. That trial is expected to begin March 4, in Washington, D.C., but the lawyers for Trump have asked the district court to pause the proceedings while he pursues his appeal.

The journalists are quite enthusiastic, as Smith moved to "bypass the Court of Appeals", and go directly to the Supreme Court, for a decision on a "fundamental issue". He wants the Supreme Court to issue a ruling immediately, on whether Trump is immune from criminal prosecution. Trump maintains that a sitting President is above the law, so that everything he did, while President, including the "events of January 6", was legal.

In response, the Supreme Court agreed to "speed up consideration" of Smith's request, and directed the lawyer's of Trump, to submit a response to the request by December 20.

This is their idea of working at lightning speed!

Smith has also asked the Supreme Court to consider Trump's argument that he is "constitutionally protected from prosecution", because he was impeached by the House of Representatives, and acquitted by the Senate, while charged with similar crimes.

It is clear that Jack Smith is opposing the plan of Trump and the billionaires, of setting up a fascist society. It is also very likely that the justices of the Supreme Court are also doing their part, to combat this fascist threat.

This is significant, as it means that the middle class is joining the working class, in the revolutionary movement. I say this because the lawyers for the prosecution, such as Jack Smith, and all justices on the Supreme Court, are members of the middle class.

While significant, it is also strange that those same middle class legal experts, do not go right to the heart of the matter. This is to say that they should go straight to the Supreme Court, and challenge the validity of the 2020 federal election, on the grounds that it did not follow the procedure laid out in the Twelfth Amendment. At the same time, they should request that the Supreme Court order the upcoming 2024 federal election, follow that Constitutional procedure.

The fact that these experts on Constitutional law are *not* doing this, is perhaps because they are still trying to preserve the "Two Party System"! At the same time, they are trying to oppose the plan of the billionaires, to establish a fascist regime, with Trump as the figurehead dictator. If that is the case, they are "living in denial"!

To such middle class people, I have two words of advice. *Wake Up!* Your middle class days are numbered! The ruling class of multi billionaires, the bourgeoisie, have recently declared war on the middle class! All businesses but five are "Too Small To Succeed"! The assets of all of those businesses -Yours!- are about to be picked up by the billionaires! You are about to be ruined! Forced into the ranks of the proletariat! The billionaires have already decided to force all middle class people into bankruptcy! Now is not the time for "half measures"!

They have also decided to change their method of rule. The "Two Party System" is not working, so they have decided to establish a fascist society.

The attempts of well meaning, middle class intellectuals, to oppose this fascist society, while preserving the Two Party System, are somewhat touching, but hopeless. Not about to happen!

Allow me to stress the fact that your standard of living, is about to nose dive! Through no fault of your own! The billionaires have declared war on you! Cut your losses! Join the working class, the proletariat, in the war against the billionaires! In the war for Scientific Socialism! Revolution and the subsequent Dictatorship of the Proletariat! You have no future under monopoly capitalism! You have a bright future under Scientific Socialism!

Only the working class, the proletariat, can succeed in overthrowing the ruling class of billionaires. The only way that can happen, is through revolution. We need to raise their level of awareness, to the level of true Marxists, Communists, those who call for Scientific Socialist Revolution! The Dictatorship of the Proletariat!

On that subject, I am reminded of a statement someone made, many years ago. He was an actor, who became supremely successful, a "Movie Star", a "celebrity". He said that as soon as someone becomes a celebrity, two very strange things happen. The first strange thing, is that people ask these celebrities questions, concerning a subject upon which they know nothing. The second strange thing is that these celebrities answer those questions!

I would add, that a third strange thing happens, the strangest thing of all! People listen to those celebrities! Especially the "Movie Stars"!

No one can explain this, but it does help to explain the reason that Trump is so popular! He is a celebrity! A Movie Star!

I can only suggest that the lawyers continue to fight Trump and the billionaires, in court and outside court, against fascism. At the same time, use celebrities, preferably "Movie Stars", to explain that which is happening.

Professional middle class writers can prepare the script, and professional middle class entertainers can prepare the presentation. The more entertaining the video, the more beautiful and popular the actors, preferably Movie Stars, the better! Their fans will pay strict attention to that which they say! Watch how fast the working people abandon Trump and his calls for fascism, and embrace the Movie Stars! Feel free to call for Scientific Socialism, Revolution and the subsequent Dictatorship of the Proletariat! They will listen to Movie Stars!

In previous articles, I have documented other ways to fight this plan of the billionaires, to set up a fascist society, through Councils and training working class people, in preparation for the revolution. There is no need to repeat it here.

Also as previously mentioned, now is the time to form a true Communist Party, Dictatorship of the Proletariat. Another job for middle class intellectuals, urgently required.

The alternative is the ruin of the middle class, the end of the democratic republic, and subsequent fascist rule, led by Trump.

CHAPTER 15

TRUMP MORE OPENLY FASCIST

The journalists who work for the main stream press, are referring to the current political situation, as a "weird moment in America". That is one way of putting it! I would argue that a more accurate description is that of *dangerous!*

Without doubt, it is clear that Trump is becoming ever more openly fascist. It is true that he is "echoing Hitler and Mussolini", referring to political opponents as "vermin". He is also raising the spectre of "hordes of illegal immigrants", pouring across the border and "poisoning the blood" of Americans. That apparently is a direct quote from Hitlers book, Mein Kampf! His proposed solution is to build huge detention camps, in preparation for deporting millions of people!

This is not to imply that Trump has read Mein Kampf, as that is a book, and as is well known, Trump does not read books. He merely reads from a teleprompter! Those who write his speeches, referred to as his "handlers", are the people who read books, especially those on fascism. They are now becoming ever more brazen, openly using the language which was so successful, in placing Hitler and Mussolini in power.

The journalists, who work for the main stream press, are deeply concerned, properly so. Yet their understanding of the problem is flawed, possibly because they either cannot, or will not, think in class terms. Instead, they refer to "groups of people" that are being targeted by Trump, amid his growing "authoritarianism". It is not authoritarianism! It is fascism!

Those journalists cannot seem to understand that it is the *class* of *monopoly capitalists,* the *billionaires,* the *bourgeoisie,* who have decided to change their method of rule, over the *class* of *workers,* the *proletariat.* The "Two Party System" is no longer working, so they plan to replace that system. The new method of rule is known as fascism. Their plan is to place a "figure head" dictator in charge of this new fascist society. That figure head dictator is Donald Trump.

Yet it would appear that the billionaires are not entirely united, in their decision to set up an American fascist government. One member of the ruling class of billionaires, described as a "GOP Mega donor", has just announced that he is supporting Nikki Haley, as the Republican candidate for the presidency.

Those who contribute millions towards the Republican Party, are referred to as "GOP Mega donors". Those "donations" come at a price! Candidates who accept those "donations", and are elected to office, end up "in the pocket" of those same billionaires! Bought and paid for!

That is not the only indication that the ruling class of monopoly capitalists, the billionaires, are divided. The state of Colorado is "making headlines", in that the State Supreme Court just ruled that Trump cannot run for president, within the state of Colorado. His name cannot appear on the ballot, as the proposed candidate for the presidency, on behalf of the Republican Party!

That decision is based on Section 3 of the Fourteenth Amendment, which states that no one can run for office who "has engaged in insurrection or rebellion" against the United States. The Court has ruled that Trump (allegedly) took part in the "January 6 insurrection".

The implications are "far reaching"! If this decision stands up in Colorado, it could spread to numerous other states! His chances of winning the Republican nomination for presidency, just "nose dived"!

Of course the "Trump team", the lawyers who work for Trump, can be expected to appeal this decision, to the United States Supreme Court. The sooner the better! If the Court chooses to consider the case, and then rules in favour of the Colorado State Supreme Court, then that ruling could apply to all states. Trump could effectively be "off the ballot", all across the country! That is up to the Supreme Court of the United States! Their decision!

As for those who maintain that the Supreme Court of the United States, SCOTUS, is completely impartial, above partisan politics, feel free to face the fact that we live in a class society! The billionaires are in charge! They rule! All government officials, including judges, "dance to their tune"!

As for those who are skeptical, may I suggest you consider the fact that for the last *twenty years,* a Supreme Court Justice has been accepting "gifts" from a billionaire. These gifts are nothing other than *bribes!* The value of these bribes are in the millions! That places this Supreme Court Justice "in the pocket" of a billionaire! Bought and paid for! He is not the only one!

Granted, by law, all Justices are required to report all gifts over the value of $415. This never happens! The law is never enforced! The law cannot be enforced! There is no punishment for breaking that law! The Justices interpret the law as they see fit! The only possible recourse is to be impeached by Congress. That is not about to happen!

The anticipated Supreme Court decision should help to reveal the divisions within the ruling class of billionaires. We can only hope that those divisions are deep and irreparable.

Trump is well aware of the implications of that potential Supreme Court decision. He is also well aware that his former personal lawyer was involved in a civil law suit. The jury awarded damages in the amount of

$148 Million! The lawyer responded by doing the only thing he could do. He filed for bankruptcy.

As I consider this topic to be so important, I will go into it in more detail. This particular -former!- lawyer for Trump was also -formerly!- referred to as "America's mayor". Very popular! He was also once the Time magazine "Person of the Year". Now that is high praise! Such is no longer the case! "How the mighty have fallen"!

The reason I consider this to be so significant, is because of the manner in which certain journalists are covering the story. In particular, I was impressed by the presentation that was given on an American "news commentary show", one which bills itself as "progressive". They make no claim to be socialist or even Leftist, although they have been so accused. I have tried to re-create the speech given, as accurately as possible. Bear in mind, that it was not so much the words that this commentator used, but the manner in which she spoke, which deeply impressed me.

In my opinion, she spoke out of a sense of deep bitterness and frustration, when she referred to that lawyer as possibly the "dumbest man in America, and I do not say that lightly". She went on to say that even though he was just ordered to pay $148 Million for defaming two women, "he still cannot shut up", but is still "whining and defaming women".

Her summary I found to be quite touching: "You hit a point in your life when you realize that all of the leaders that were touted in this country are the most incompetent buffoons imaginable."

Absolutely correct! She "hit the nail right on the head"! If she was not middle class, I would say that she is starting to become class conscious. But as all middle class people are class conscious, aware of themselves as a class, perhaps it is more correct to say that she is becoming aware of the fact that she has been lied to, all her life, by those for whom she had the utmost respect. A painful lesson!

The fact is that the billionaires, as well as their loyal and devoted servants, are habitual liars. It is just their nature to lie! They cannot stop lying! What is more, those same liars write the history books. It follows that those same history books are filled with distortions, and outright lies! Now she knows!

It is reasonable to assume that there are countless middle class intellectual people, who are of the same opinion. No doubt, they too are feeling bitter and frustrated. With good reason, I might add.

It is also reasonable to assume that this may just be enough to "force them further to the Left", so to speak. It may even inspire them to read the Essential Works of Marx and Lenin, once again, but this time with an open mind.

I say this because all middle class intellectuals have attended University, and have been "exposed" to the Scientific Socialist theories of Marx and Lenin. Bear in mind that only the distortions of those theories, are taught in University.

There is an urgent need for those intellectuals to take the next step, to become true Scientific Socialists, to call for revolution and the subsequent Dictatorship of the Proletariat.

I use the work "urgent", if only because the billionaires plan to set up a fascist society, with Trump as the figurehead dictator.

Now to return to the implications this huge lawsuit, against the former attorney of Trump, has for Trump.

Trump can only imagine the amount of money he will be forced to pay, in his civil law suits. The situation just became far more acute! This calls for a little explanation.

At the risk of offending a great many of my readers -which cannot be helped- allow me to state that all hunters are well aware that a wounded predator is supremely dangerous. Especially if that animal is cornered!

The only thing standing between Trump and prison -he is facing 91 felony charges!- is his planned occupation of the White House. The same is true of Trump and financial ruin. Either political power or prison and bankruptcy! If he succeeds in becoming President, once again, only this time as a fascist dictator, then he will not have to worry about any criminal or civil charges! Dictators do not worry about such little details!

On the other hand, if, on the "civil fraud front", so to speak, the judge in New York City were to hit him with a huge fine, and bar him from doing business in New York state, that could "throw a monkey wrench in the works". This could happen as early as January.

Then there are the criminal charges. The first trial is scheduled to begin as early as March 4. If convicted, he could be sentenced to a lengthy prison term.

As for those who may reasonably object that he would merely appeal such court decisions, I can only respond that such appeals cost money. Lots of money. If his businesses are crippled, that is out of the question.

The closest followers of Trump, including his lawyers, are in it strictly for the money. If the money runs out, then they run out. Like rats from a sinking ship! Forget the appeals!

Facing such a threat, Trump may act as a wounded, cornered predatory animal. He may lash out in any direction. He may not wait for the fall elections, to "make his move". As he said, he plans to become dictator, "*at or before*" the next federal elections. He was not joking! He and his followers may attempt to stage a coup, and set up a fascist dictatorship, at any time.

This is to stress to my Leftist intellectual, middle class colleagues, the importance of becoming active. Capitalism is in its death throes! As that is the case, we can expect Trump and the other billionaires to lash out wildly! No one knows what to expect! We just know what they have planned! A fascist society!

The necessity of raising the level of awareness, of the working class, is ever more urgent. The less advanced working people, or "MAGA supporters", have merely been misled. They are not the enemy! Yet the fascists cannot succeed without their support!

Until quite recently, efforts to raise their level of awareness, was restricted to the production of quite popular literature. Happily, advances in technology have made possible another "avenue of approach". Of course I am referring to videos.

As mentioned in previous articles, working people pay strict attention to the opinion of "celebrities". Especially "Hollywood celebrities", or "Movie Stars", if you prefer. Now is the time for all Leftist people, especially those involved in the entertainment industry, to put aside their differences, and focus on stopping the billionaires, in their attempt to establish a fascist society, with Trump as the dictator.

We do not have to love each other! A temporary alliance is perfectly acceptable. We merely need professional writers to come up with a script, written in popular terms that common people can understand. The more entertaining the better.

Then, those who are famous "Movie Stars" can make the proper delivery, reading from a teleprompter. Such people do not necessarily have to agree with everything they are saying. But then, they are probably used to that! After all, there is a reason it is called acting!

Our goal is to make the working class, the proletariat, aware of themselves, as a class. As opposed to the class of monopoly capitalists, the billionaires, the bourgeoisie. The wealth of the billionaires, comes at our expense. The more poverty stricken we are, the more wealthy the billionaires. For that reason, we are class enemies.

The class of monopoly capitalists, the billionaires, have to be overthrown. This can happen only through revolution. Then we have to establish a society of Scientific Socialism. This means smashing the existing state apparatus, and setting up a working class state apparatus, in order to crush the billionaires, as they make every effort to "restore their paradise lost". This new state apparatus is known as the Dictatorship of the Proletariat.

Granted, to put that in video form is a "tall order", but then, perhaps no more difficult than putting it in written form. Further, under far more difficult circumstances, it has previously been accomplished.

Bear in mind that the fans of these "Movie Stars" will pay strict attention to whatever their heroes say! Strange but true! We can expect the "MAGA" people to desert Trump in droves, and flock to Scientific Socialism, and embrace the Dictatorship of the Proletariat.

Perhaps all middle class intellectuals, both Left and Right, can take some inspiration from the fact that the billionaires have decided to wipe out the middle class. As countless banks and businesses fail, those who own shares in those enterprises, including stocks and bonds, will lose everything. All will be forced into the ranks of the proletariat.

Not that all middle class intellectuals are about to see the "writing on the wall"! We can expect the most hard core Right wing souls to continue to live in denial. Even when their world comes crashing down around them, as it most certainly will, they may continue to defend their lords and masters, the billionaires.

By contrast, within the entertainment industry, those Leftist middle class intellectuals who support the revolutionary movement, will be well rewarded, after the revolution, under Scientific Socialism, in the form of the Dictatorship of the Proletariat.

CHAPTER 16

CREATE AN INDEPENDENT ANTI FASCIST SOCIETY

The Supreme Court has just responded to the request of the Special Prosecutor, Jack Smith, to "expedite" the ruling on whether or not Trump has any immunity, concerning the alleged crimes he committed on January 6, 2021, at the time of the so called "insurrection" in Washington.

This is to say that Smith was trying to bypass the D.C. Court of Appeals. That Court is scheduled to hear "oral arguments" on the matter, in early January. Regardless of the ruling, it will no doubt be appealed to the Supreme Court, so such a request is completely reasonable.

The Supreme Court responded with a simple one sentence statement of denial, without giving any reason. Chalk one up for Trump! He has succeeded in delaying that trial, yet again! Thanks to the Supreme Court, which is now being referred to as "The Dancing Court"! In the finest tradition of the bourgeois, they are once again, "passing the buck". Not making any decisions! Or at least delaying any decisions! This is also referred to as "kicking the can down the road"!

The journalists, who work for the mainstream press, are resorting to "verbal gymnastics", in an attempt to report the news, without offending those for whom they work, the billionaires. They are referring to this court case as a "legal landmine", in a country that is "deeply divided", involving a "critical legal question", one which is "so fraught with politics", to be decided by a "deeply divided court". Hogwash!

The "deeply divided country", to which they refer, is nothing other than a country divided by *class conflict*. It is now so fierce, it is approaching *open class warfare!* The *working class*, the *proletariat*, versus the *monopoly capitalist class*, the *billionaires*, the *bourgeoisie!* Further, the "legal landmine" is not a "critical legal question", nor is it "fraught with politics". It is a matter of *Constitutional law*, to be decided upon by the Supreme Court! Whether or not that Court is also "deeply divided", is not the issue! It is their *duty!* It is just that simple!

The fact that the corruption, which is characteristic of all capitalist countries, has spread to all branches of the government, including the Supreme Court, does not change that fact.

Now this completely corrupt class of monopoly capitalists, the billionaires, are determined to establish a fascist society. They must be stopped! It is not reasonable to expect the corrupt state apparatus, which is in the service of the billionaires, to oppose the plans of those same billionaires, to establish a fascist society.

Yet outside the government, the opposition to this proposed state of fascism is broad and deep. With good reason! It is characteristic of fascists, to persecute certain minorities. This persecution may be based upon race, religion or ethnic background. Even great wealth may not save them! As well, anyone whom they perceive to be "enemies", is "fair game"! "Vermin"! To be "liquidated"!

For that reason, an anti fascist coalition should not be strictly "Leftist". It should include all of those who believe in a democratic republic. Such people may, or may not, be a member of one of the two mainstream political parties, Democratic or Republican.

In the interest of preserving our democratic republic, it is necessary for those who consider themselves to be "Leftists", including anarchists, socialists, democratic socialists, independent socialists and Communists, among others, to work with those who are considered to be "Ultra Right Wing".

As for those who consider themselves to be "Ultra Left Wing", and object to working with such disagreeable people, may I suggest that there are times when such *temporary* alliances, with those who are highly unreliable, are necessary! According to Lenin!

Truly, "politics makes for strange bedfellows"! Besides, bear in mind that they hate us, just as much as we hate them!

The only qualification, is that any Communist who works within such a coalition, must be free to express those Communist beliefs. That includes the necessity of Scientific Socialism, through revolution, and the subsequent Dictatorship of the Proletariat. Without such an understanding, any such coalition, is out of the question.

This proposed coalition should take the form of a Society, one which is non profit, focused only on opposing fascism. It should have no opinion on any outside issues. It should neither support, nor oppose, any causes. The one and only goal, of that Society, should be that of opposition to fascism.

There are various ways to oppose fascism. One of these methods involves court challenges. One of those court challenges, which I am proposing, is considered to be "forbidden"! That is to challenge the "sacred", "Two Party System"! Sacred or not, it is Unconstitutional!

It may well be objected, that the American Two Party System, is thought to be, "at the heart of our democratic republic". That may well be. That is not the issue. At issue is the law. That law is referred to as the Constitution.

It may further be objected, that I am not an expert on Constitutional law. True! Guilty as charged! I am not even a lawyer! I am merely expressing my opinion. Not that my opinion matters. The one and only opinion that matters, is the opinion of the Supreme Court! They have never expressed an opinion on this subject, because the Two Party System has never been challenged!

As I have mentioned in previous articles, the Twelfth Amendment lays out the procedure, to be followed, for all *federal* elections! I have deliberately stressed the word "federal", as it is now the custom to refer to this as a "presidential" election. It is *not* presidential. It is *federal!* As it is so important, I have chosen to reproduce it, in full:

Twelfth Amendment

The Electors shall meet in their respective states and vote by ballot for President and Vice-President, one of whom, at least, shall not be an inhabitant of the same state with themselves; they shall name in their ballots the person voted for as President, and in distinct ballots the person voted for as Vice-President, and they shall make distinct lists of all persons voted for as President, and of all persons voted for as Vice-President, and of the number of votes for each, which lists they shall sign and certify, and transmit sealed to the seat of the government of the United States, directed to the President of the Senate;–the President of the Senate shall, in the presence of the Senate and House of Representatives, open all the certificates and the votes shall then be counted;–The person having the greatest number of votes for President, shall be the President, if such number be a majority of the whole number of Electors appointed; and if no person have such majority, then from the persons having the highest numbers not exceeding three on the list of those voted for as President, the House of Representatives shall choose immediately, by ballot, the President. But in choosing the President, the votes shall be taken by states, the representation from each state having one vote; a quorum for this purpose shall consist of a member or members from two-thirds of the states, and a majority of all the states shall be necessary to a choice. [And if the House of Representatives shall not choose a President whenever the right of choice shall devolve upon them, before the fourth day of March next following, then the Vice-President shall act as President, as in case of the death or other constitutional disability of the President.–]The person having the greatest number of votes as Vice-President, shall be the Vice-President, if such number be a majority of the whole number of Electors appointed, and if no person have a majority, then from the two highest numbers on the list, the Senate shall choose the Vice-President; a quorum for the purpose shall consist of two-thirds of the whole number of Senators, and a majority of the whole number shall be necessary to a choice. But no person constitutionally ineligible to the office of President shall be eligible to that of Vice-President of the United States.

Clearly, there is no mention of any "Two Parties"! Nor is there any mention of any "Popular Vote", "District", or "Running Mate"!

The argument can also be made that the states do not have the right to "meddle" in any federal election. This is to say that forcing the Electors, by state law, to vote for a particular candidate for the Presidency, or of the Vice Presidency, of either the Republican Party, or of the Democratic Party, is Unconstitutional!

There is a reason that *no* federal election has ever been challenged in court. Because the experts on Constitutional law are well aware, that the Two Party System is Unconstitutional! They do not want to "open that can of worms"! To challenge the 2020 federal election, on the grounds that it did not follow the procedures laid out in the Constitution, would almost certainly result in the Supreme Court ruling in their favour. Effectively, such a ruling would establish the fact that Biden is a fraudulent President, and Harris is a fraudulent Vice President!

Such a Supreme Court decision would also mean that *all* federal elections, dating back to the days of the Civil War, were illegal! *All* of those Presidents, and Vice Presidents, were *fraudulent!*

Further, it would also mean that the upcoming federal election of 2024, would have to follow the procedures laid out in the Constitution. No popular vote! Any such vote would be meaningless! Only the states can appoint Electors, and all state laws, requiring those Electors to vote for the candidates of a major political party, would be struck down as Unconstitutional!

For that reason, it would not matter if Trump was the candidate of the Republican Party, or not. Each and every Elector would be able to vote for the individuals of their choice! Such individuals may, or may not, be the candidates of a particular party! As per the Constitution!

That is the anticipated result of such a Supreme Court challenge, to the Two Party System. Such a ruling, that the Two Party System is Unconstitutional, would "throw a monkey wrench" into the plans of the billionaires, to set up a fascist society, with Trump as the dictator.

May I suggest, to the most avidly "Right Wing" anti fascist people, that our options are limited. We can stop the ruling class of billionaires, from setting up a fascist state, with Trump as the dictator, or we can do nothing. Doing nothing is not an option!

Bear in mind that Trump has made it clear that *all* of his opponents are *vermin!* To be placed in huge concentration camps! To start! Guess what happens to all such vermin after that! Expect a *Second Holocaust!* That is the plan of Trump!

As well, no doubt such an Anti Fascist Society will attract numerous professional people, including "celebrities". It is such celebrities, especially those who are seen as "Movie Stars", who have great power! Spiritual power! Possibly far more than they realize! For reasons which defy any rational explanation, working people pay strict attention to anything they say! To such people, I say, use that power! To oppose Trump and the fascists!

My suggestion is that professional writers can prepare a script, to be read by those considered to be Movie Stars -the more popular and beautiful the better!- recorded in video form, and then placed on the internet. Better yet, a series of videos, written in a popular, entertaining manner. Watch how fast they go "viral"!

These videos do not have to be "Leftist". They have to be anti fascist. We can expect their effect on the common people, to be immediate and dramatic. Watch how fast working class people abandon Trump and the fascists, and embrace the "Movie Stars", along with a democratic republic!

Bear in mind that working class people are avid readers. They also pay strict attention to the news. So the videos should make reference to current events, while being presented in an entertaining manner, without being condescending.

The success of these videos, will further enhance the careers of all involved in their production.

By contrast, the success of Trump and the fascists, is not to be considered.

CHAPTER 17

CONCERNING LENIN'S LETTER TO AMERICAN WORKERS

Now that the year is coming to a close, it is customary for the journalists to summarize the major events that happened, in the previous year, or at least the major political events. Then they tend to predict the political events, in the forthcoming year. As if they had the foggiest clue!

Yet one guest has distinguished herself, as she provided an analysis which is close to the truth. This is all the more remarkable, in that she was able to point out the class distinctions, without once mentioning the word class!

She was introduced as a "political reporter", and a "former Republican strategist and pollster". She is convinced that most Americans are of the opinion that "voters admire the fighter mentality", that they are looking for someone "to fight against a system that seem to be working against them", because it is a system that "works for the elite, for the few, and not for the people, not for them".

My dear "former Republican strategist", you are so right when you suggest that Americans "admire the fighter mentality"! No one ever suggested that Americans are a bunch of wimps! Pacifists, they are not! Your country was borne through revolution! It was the Revolutionary War against British colonialism, which gave birth to your country! Further, it was the American Civil War which destroyed the completely reactionary class of slave owners! Americans have a proud history of revolution!

You are also correct when you state that "voters", or at least *working class* voters, are looking for "someone", by whom you mean a *leader(!)*, who is going to "fight against a system that seems to be working against them".

It only "seems" that way, because that is precisely the case! That "system", to which you refer, is nothing other than the bourgeois state apparatus, which has been set up, with the express purpose of *crushing* the working class! The working class knows this, with their class instincts! They know that the "system", truly "works for the elite", those who were formerly referred to as the "1%", now referred to as the billionaires, technically referred to as the bourgeoisie. That is the *class* of people, whom you, ever so politely, refer to as the "elite".

Dear lady, try to get it through that thick, middle class skull of yours, the fact that classes exist! What is more, since the time they came into existence, they have been in conflict! Class conflict! The capitalist class, the bourgeoisie, against the working class, the proletariat! Our interests are diametrically opposed. That which is in the best interest of one class, is in the worst interest of the other class.

No doubt, you are aware of the fact, that the revolutionary movement is becoming ever stronger. The "class conflict" is approaching open "class warfare". Perhaps you think that, as a member of the middle class, the petty bourgeois, you are in a position of "safe neutrality". Such is hardly the case! There is no such "position"!

The billionaires recently declared war on the middle class! They plan to wipe out that class! All banks but eight, and all businesses but five, are *Too Small To Succeed!* As they collapse, the billionaires will pick up their "assets". At the same time, the middle class people who have money in the bank, as well as shares in those businesses, stocks and bonds, are about to lose everything.

Remarkably enough, our current situation was anticipated by Marx and Engels. In the Communist Manifesto, they state: "Our epoch, the epoch of the bourgeoisie, possesses, however, this distinctive feature: it has simplified the class antagonisms. Society as a whole is more and more splitting up into two great hostile camps, into two great classes directly facing each other: bourgeois and proletariat."

As for the American working class, it may come as a complete surprise, to learn that Lenin had the utmost respect for revolutionary Americans! In fact, he wrote a letter to American workers, on August 22, 1918, less that a year after the Great Russian October Socialist Revolution. I have decided to reproduce parts of it here:

"The history of modern, civilized America opened with one of those great, really liberating, really revolutionary wars of which there have been so few....That was the war American people waged against the British robbers who oppressed America and held her in colonial slavery....since then, bourgeois civilizations has borne all its luxurious fruits. America has taken first place among the free and educated nations in level of development of the productive forces...At the same time, America has become one of the foremost countries in regard to the depth of the abyss which lies between the handful of arrogant multimillionaires who wallow in filth and luxury, and the millions of working people who constantly live on the verge of pauperism. The American people, who set the world an example in waging a revolutionary war against feudal slavery, now find themselves in the latest, capitalist stage of wage-slavery to a handful of multimillionaires, and find themselves playing the role of hired thugs.... The American people have a revolutionary tradition that has been adopted by the best representatives of the American proletariat, who have repeatedly expressed their solidarity with us Bolsheviks.... When we are confronted with the vastly greater task of overthrowing capitalist *wage* slavery, of overthrowing the rule of the bourgeoisie -now the representatives and defenders of the bourgeoisie, and also the reformist socialists who have been frightened by the bourgeoisie and are shunning the revolution, cannot and do not want to understand that civil war is legitimate and necessary.

"We know that fierce resistance to the socialist revolution on the part of the bourgeoisie is inevitable in all countries, and that this resistance will *grow* with the growth of this revolution. The proletariat will crush this resistance; during the struggle against the resisting bourgeoisie, it will finally mature for victory and for power....The revolution is developing in different countries in different forms and in different tempos (and it cannot be otherwise)....We are banking on the inevitability of the world revolution...and we know that revolutions are not made to order, or by agreement....and that *before* the world revolution breaks out, a number of separate revolutions may be defeated. ...Slowly but surely the workers are adopting Communist, and Bolshevik tactics and marching towards the proletarian revolution, which alone is capable of saving dying culture and dying mankind.

"In short, we are invincible, because the world proletarian revolution is invincible." (italics by Lenin)

The full letter is available on the internet, and all Americans, working class as well as middle class, are encouraged to read it.

Now to return to our current situation. The billionaires can no longer rule in the old way, that of the democratic republic, and have decided to change their method of rule. The new method of rule they have decided upon, is that of fascism, with Trump as the figurehead dictator.

This begs the question: How can they be stopped?

The Communist Manifesto provides us with the answer: "But not only has the bourgeoisie forged the weapons that bring death to itself; it has also called into existence the men who are to wield those weapons -the modern working class -the proletarians."

Marx made it clear, that the proletariat is the class which is destined to overthrow the capitalist class of billionaires. No other class can manage this.

Yet one major problem is that the working class, the proletariat, is not aware of itself as a class, with its own class interests. The conditions of life, of the working class, do not lead to that awareness.

This gives rise to considerable confusion, within the class conflict, which I can compare only to a boxing match, in which one boxer is blind folded. Of course, in my imaginary boxing match, the proletariat is the one who is blind folded, while the capitalists, the billionaires, are not.

As can be well imagined, the boxer who is blind folded strikes out in all directions, and on occasion, is able to land a lucky blow. By contrast, his opponent, who can see quite well, is able to land one blow after another.

This is admittedly a gross oversimplification, but it serves to drive home the point that the proletariat is at a distinct disadvantage, as it is not class conscious. The level of awareness, of the proletariat, must be raised to the level of Scientific Socialists.

This is to say that they must become aware -better yet, embrace!- the revolutionary theories of Marx and Lenin. They must become aware that it is necessary to overthrow the ruling class of monopoly capitalists, the multi billionaires, and to smash the existing state apparatus, that which has been set up by the billionaires, as a means of crushing the working class. A new state apparatus must be set up, to crush the "desperate and determined resistance of the billionaires", as they make every effort to "restore their paradise lost". This new state apparatus is referred to as the Dictatorship of the Proletariat. True Scientific Socialism!

Yet how are they to become aware of this? The revolutionary theories of Marx and Lenin are taught -more accurately, distorted!- only in University. Very few working class people are able to attend University. The Communist Manifesto provides us with the answer:

"Entire sections of the ruling classes are, by the advance of industry, precipitated into the proletariat, or at least threatened in their conditions of existence. These also supply the proletariat with fresh elements of enlightenment and progress".

I would suggest that this has been taking place for some considerable time now, and is accelerating. Recently, countless upper class people have been ruined, forced into bankruptcy and the ranks of the proletariat. They bring with them their awareness of the existence of classes, the class conflict, and the revolutionary theories of Marx and Lenin.

I would further suggest that the class struggle is approaching open class warfare. Or as the Communist Manifesto states:

"Finally, in times when the class struggle nears the decisive hour, the process of dissolution going on within the ruling class, in fact within the whole range of old society, assumes such a violent, glaring character, that a small section of the ruling class cuts itself adrift, and joins the revolutionary class, the class that holds the future in its hands".

We are nearing the "decisive hour". Not all members of the "ruling class", the billionaires, are determined to establish a fascist republic! Some of them are opposed to this, with good reason! Under a state of fascism, even certain billionaires may be persecuted! Based on religious or ethnic backgrounds! Or because they are intellectuals! It only makes sense for them -survival!- to "cut themselves adrift", and "join the revolutionary class", the proletariat.

In conclusion, I can only suggest that American workers, and middle class intellectuals, honour the memory of Lenin, by living up to his every expectation. Follow in the footsteps of your revolutionary ancestors!

The most important thing now, the "key link", is to raise the level of awareness of the proletariat, in preparation for revolution, and Scientific Socialism, in the form of the Dictatorship of the Proletariat. All actions we take now, must be made, with that goal in mind. This calls for a "multi pronged" approach, involving various methods, some of which are legal, as well as others, which are something other than legal.

Use the judicial system! The very judicial system that the capitalists have set up to crush the lower classes, can be used against the capitalists! Experts on Constitutional law should challenge the Two Party System, on the grounds that it is Unconstitutional. It violates the Twelfth

Amendment! A possible Supreme Court ruling, in our favour, can only be a bonus! The main thing is to expose the hypocrisy of the capitalist system.

Also on the "legal front", encourage one and all to join the two mainstream political parties, preferably as card carrying members. The leaders of those common people can then run for any and all political offices. The membership of any particular political party, or a third party, is a matter of no consequence. Just be sure not to run against any "Member of The Squad", or any "Bernie Bro".

Allow me to stress the fact that our goal is to raise the level of awareness of the proletariat, to prepare them for the Dictatorship of the Proletariat. In the process of running for office, of trying to "change the system from within", working people will learn that the state apparatus has been set up to support the billionaires. In this way, they will become more class conscious.

Establish Councils and train the members for warfare! Arm and equip those workers! Prepare them for the insurrection! The more training they receive now, the more likely it is that the insurrection will not only succeed, but also be almost bloodless! As was the case of the Russian Great October Socialist Revolution!

Form a true Communist Party, Dictatorship of the Proletariat! This is a task for middle class intellectuals, perhaps with the assistance of advanced workers, as it is clearly beyond the ability of the proletariat.

Reach out to the less advanced workers, those who have been misled, generally referred to as "MAGA" people! Prepare popular literature, and especially videos, for those workers! Whenever possible, have "celebrities", especially young and beautiful "Movie Stars", explain the class struggle, in terms they can understand. Such people listen to those Movie Stars! Be forceful! Bear in mind, that Americans "admire the fighter mentality"! Do not "pull your punches"!

Above all else, remember that, as Lenin said, "We are invincible, because the world proletarian revolution is invincible".

CHAPTER 18

AN AMERICAN FASCIST: HENRY FORD

hanks to Donald Trump, people are now openly discussing fascism. The reason for this is quite simple. It is because Trump is a fascist. He plans to become president, once again, and set himself up as dictator, in charge of a fascist state.

As this has happened before, in Italy and Germany, it is feared that it could happen in America. These fears are well grounded. In fact, Trump plans to come to power, just as Hitler came to power in Germany, one way or another.

In 1923, Hitler made an attempt to come to power, the "other way". He tried to overthrow the German government. This has gone down in history as the "Beer Hall Putsch". The attempt was unsuccessful, and Hitler was sent to prison. While in prison, he wrote his famous book, "Mein Kampf", or "My Struggle". It became the "bible" of National Socialism, or "Nazism", in the German Third Reich.

This Nazi manifesto reveals the racist ideology of the fascists. It identifies the "Aryan" as the master race. It also singles out the "Jew" as the "parasite", responsible for the defeat of Germany in the First World War. As well, it stresses the need for Germans to seek "Lebensraum", or "living space", in the East, at the expense of the Slavs. This is justified through hatred of the Marxists, especially the "Bolsheviks", of the Soviet Union.

It is to the credit of American journalists, that they have documented the fact that the rhetoric of Trump, is taken directly from the manifesto of Hitler. Yet there is more to it than that.

The first edition of Mein Kampf was clearly written by Hitler alone. Critics have described it as "repetitious, wandering, illogical and filled with grammatical errors", that which is characteristic of a "half educated man". Yet those same critics say that the book was "skillfully demagogic", appealing to the "dissatisfied elements in Germany". Of course, it was also anti semitic, anti democratic and anti Marxist.

In popular terms, we can say that Hitler was "talented". He had a certain "charisma", which is difficult to define. It is not something which is taught in University. Those who have charisma, tend to possess a certain superficial charm, as well as an animal like "cunning". He instinctively learned, "on the streets", how to take advantage of the most basic impulses of the lowest strata of society. In fact, in terms of manipulating people, he was supremely successful.

This made Hitler a leader, especially among the "less advanced", lower class members of society. Such people are politely referred to as "social outcasts". Less polite terms, but more common, include such labels as "ragtag and bobtail", "lowlife", "gutter sweepings", "trailer court trash", "dregs of humanity", "scum of the earth" and "great unwashed mob". The scientific term is "lumpen proletariat".

Now in America, we have another demagogue, no less charismatic, no less cunning, no less stupid, no less determined to establish a fascist society, with himself as dictator. That man is Donald Trump. He too, is no less dangerous than Hitler!

That being said, this begs the question: How could Hitler, a man of limited education, write such a book? Despite its numerous flaws, it became wildly popular. The short answer is that Hitler did not write

Mein Kampf. He merely plagiarized the work of a famous American industrialist, Henry Ford!

Henry Ford is best known as a supremely successful American businessman, the proud owner of Ford Motor Company. He was a multi millionaire, and has been given credit for creating the first moving assembly line. This process is now used in large scale manufacturing, and has resulted in the production of vehicles at a record breaking rate. He is less well known as a dedicated racist.

Ford was a man of action. He did not keep these racist beliefs to himself. He owned a weekly newspaper, The Dearborn Independent, and used it to spread his racist garbage.

Over a three year period, from 1920 to 1922, he wrote an impressive series of articles. These articles were later collected and published in a four volume set, titled: "The International Jew: The World's Foremost Problem".

A cursory examination of those writings, reveals that the author, Henry Ford, was a well educated man, an intellectual. He clearly devoted a great deal of time and effort, into presenting a very carefully thought out, elaborate, detailed, well organized thesis. Which in no way changes the fact that it is a racist pile of garbage! Aside from the grammar, it is no different from Mein Kampf! That is not a coincidence!

It is reasonable to assume that, while in prison, Hitler managed to get his hands on a German translation of the four volume set of articles, written by Ford. Then, with his limited academic ability, and even more limited intellect, Hitler was able to focus on the parts which he could understand, and wrote them down. These collected writings, which he copied from Henry Ford, he then published in book form. He titled that book "Mein Kampf".

This is significant, because the current American ruling class of monopoly capitalists, the billionaires, have decided to take the advice of this racist demagogue, Henry Ford. They have decided to establish a fascist state, with themselves as the "master race", and Trump as the figure head dictator.

I have chosen to use the word "significant", because they must not be allowed to establish that fascist state. They must be stopped! And, as any common person is well aware, "the best way to kill a snake, is to cut off its head"!

As the journalists have correctly pointed out, the poison which Trump is now spewing, is a direct quote from Hitler. Yet Hitler was merely quoting Henry Ford. Now it is necessary to go directly to the source, to focus on that which was written by Ford. After all, it is very likely that the American fascists are merely following the "script", written by Henry Ford.

Not that they are about to admit this! Henry Ford is considered to be an American "role model", a "hero", an "icon", an "inspiration" to countless young Americans. The "establishment" uses Ford as "proof" that in America, "anyone can make it"! Such nonsense!

It is necessary to expose Henry Ford, as the racist piece of human garbage that he was! It was Ford who provided the theoretical "justification", or the "foundation", if you will, for fascism. In Germany, it took the form of Nazis. And now the American fascists are taking the advice of Ford, attempting to establish their dictatorship here! An American fascist state!

With that in mind, may I suggest to intellectual members of the "upper classes", that there is a lesson to be learned here. Ford was clearly an intelligent man. He was not stupid! His theories, of racial superiority, were well thought out, and presented in a very scholarly manner. They were also ridiculous, completely devoid of logic! He may have been intelligent, but he was not rational!

This lack of logical, rational, clear thinking, is characteristic of all fascists. This makes them all the more dangerous, as they are completely unpredictable! Just as Hitler attempted to overthrow the German government, so too, Trump may make a similar effort, to overthrow the American government!

Still not convinced? A fine example of the lack of logic, among the fascists, is the fact that Hitler warned about the dangers of a war on two fronts. He learned this, from his experience in the First World War. Then in June of 1941, while Germany was still at war with Great Britain, he invaded the Soviet Union. War on two fronts! The very thing he had advised against, in Mein Kampf!

Fascists are completely irrational! They make no sense! Regardless of how intelligent they are! They may strike out at any group of people, for any reason, or for no reason! Because they can! And they will! Once they come to power, no one is safe from the fascists!

Now in America, the ruling class can no longer rule in the old way, and have decided to change their method of rule. The new method of rule, upon which they have decided, is fascism.

It is very likely that the intellectuals, within the class of monopoly capitalists, the billionaires, have failed to think this through. Or perhaps it is their intellectual servants, members of the middle class, who are responsible. Either way, all bourgeois intellectuals would do well to bear in mind that in Nazi Germany, all intellectuals were shot!

After the Nazis came to power, they took the advice of their intellectual advisors, those whom had helped place them in power. That advice included the *killing of all intellectuals!* So the Nazis rounded up *all* intellectuals, including the intellectuals who called for all intellectuals to be shot, and shot them!

With that in mind, my advice, to all upper class bourgeois intellectuals, is to *"bail out"!* Cut your losses! If the billionaires succeed in establishing a fascist dictatorship, as they have planned, then you can expect to be arrested and shot! This happened in Nazi Germany, and it could happen here!

The one and only alternative, is to join the "revolutionary class, the class that holds the future in its hands"! That revolutionary class is the proletariat!

The revolutionary motion is growing ever stronger! Now is the time to raise the level of awareness of the working class, to prepare them for the Dictatorship of the Proletariat!

As I have documented the manner in which this can be done, in previous articles, there is no need to repeat it here. I can only stress the fact that middle class intellectuals have a bright future, but only under Scientific Socialism, the Dictatorship of the Proletariat.

By contrast, those same intellectuals have no future under capitalism, certainly not fascism. Especially if the fascists decide to "liquidate" all intellectuals, including the intellectuals who helped place them in power. And they will! As for those who suggest that makes no sense, bear in mind that fascists tend to be not overly rational. Consider Donald Trump. Not the "sharpest tool in the shed"! Just a typical fascist.